THE IMMUTABLE LAWS
OF THE AKASHIC FIELD

ALSO BY ERVIN LASZLO

The Whispering Pond (1996)

The Consciousness Revolution—with Stanislav Grof
and Peter Russell (1999)

The Connectivity Hypothesis (2003)

Science and the Akashic Field (2004, 2007)

Science and the Reenchantment of the Cosmos (2006)

CosMos—with Jude Currivan (2008)

The Akashic Experience (2009)

The Dawn of the Akashic Age—with Kingsley L. Dennis (2013)

The Self-Actualizing Cosmos (2014)

What Is Reality? (2016)

What Is Consciousness? (2016)

The Intelligence of the Cosmos (2018)

Reconnecting to the Source (2020)

THE
IMMUTABLE LAWS
OF THE
AKASHIC FIELD

Universal Truths for a Better Life
and a Better World

ERVIN LASZLO

FOREWORD BY
MARIANNE WILLIAMSON

EPILOGUE BY
JEAN HOUSTON

WITH CONTRIBUTIONS BY
KINGSLEY L. DENNIS, MARIA SÁGI,
AND CHRISTOPHER M. BACHE

ST. MARTIN'S
ESSENTIALS
NEW YORK

First published in the United States by St. Martin's Essentials, an imprint of St. Martin's Publishing Group

THE IMMUTABLE LAWS OF THE AKASHIC FIELD. Copyright © 2021 by Ervin Laszlo. Foreword copyright © 2021 by Marianne Williamson. Epilogue copyright © 2021 by Jean Houston. All rights reserved. Printed in the United States of America. For information, address St. Martin's Publishing Group, 120 Broadway, New York, NY 10271.

www.stmartins.com

Designed by Steven Seighman

The Library of Congress Cataloging-in-Publication Data is available upon request.

ISBN 978-1-250-77384-5 (trade paperback)
ISBN 978-1-250-77385-2 (ebook)

Our books may be purchased in bulk for promotional, educational, or business use. Please contact your local bookseller or the Macmillan Corporate and Premium Sales Department at 1-800-221-7945, extension 5442, or by email at MacmillanSpecialMarkets@macmillan.com.

First Edition: 2021

10 9 8 7 6 5 4 3 2 1

CONTENTS

PART 3

GUIDANCE BY THE LAWS OF THE AKASHIC FIELD

FOREWORD

Marianne Williamson

As I write this the world is gripped by panic, beset by an unprecedented global pandemic. Fear, even hysteria, prevail in places usually immune to such oceanic emotional factors.

The greatest scientists in the world are working to find a cure, a vaccine, any successful treatment at all. Using all the tools at the command of modern medicine, doctors display acts of sacrifice and heroism in their effort to save the sick and dying.

The most unsettling part of it all? That the crisis was previously unimaginable to a modern civilization that for the most part thought it had things pretty much figured out. What science had not yet mastered, surely it would master within the next hundred years. Despite clear evidence—namely, that the twentieth century, while awash in scientific progress, was still the bloodiest in human recorded history—modern humanity continued its delusional trajectory of thinking we were so smart, so scientific, so advanced, and so economically sophisticated that instant karma never gets us.

As it turned out, we were only halfway right: it took more than an instant.

But, boy, it's got us now. While some believe the virus is a purely random event, with no deeper significance than that which is revealed by hard data, there are many with an even deeper suspicion that nature seems to be telling us something. As a friend of mine noted, "It's like a Divine Mother said to humanity, 'Go to your room. And think about what you've done.'" Isn't that the truth. Some cosmic balancing seems to be occurring now, some necessary self-correction on the part of a universe that would simply no longer tolerate our irreverence, irresponsibility, lack of humanity, and lack of compassion toward people, planet, and animals. In the words of *A Course in Miracles,* "There is a limit beyond which the Son of God cannot miscreate." Free will is one thing; self-destruction is another. We can do whatever we want to do, but ultimately we will not be saved from the effects of what we have done.

That is where this book comes in.

Do we need science to help us now? To heal diseases? To repair the earth? To create the infrastructure for a sustainable world? Of course, we do. But can science explain to us the mystery of creation, the higher purpose of it all, and the working of the universe in a way that guides us to a better world? No, science cannot. That is where spirituality comes in.

Ervin Laszlo has long been a leading voice in the evolutionary movement toward a sustainable, even thriving, human consciousness for the twenty-first century. What has now become a mainstream movement in the social sciences grew in large part out of his work over the last few decades, both with the Club of Rome and the Club of Budapest. Now, in *The Immutable Laws of the Akashic Field,* he brings to a mainstream audience some information that perhaps it took the shocking events of the current day to prepare us for. The coronavirus has rocked our world, and it could be argued that we needed to be rocked. Rocked so hard that perhaps we are ready at last to surrender some delusional prejudices that are both hallmarks and limitations of the modern worldview. Perhaps now we can open our minds, with greater understanding, to ancient wisdom now seen anew in light of its relevance to our contemporary experience.

Such spiritual insight is not antiscience. Rather, as Laszlo explains so well, truly modern science has actually prepared the way for greater spiritual understanding. And vice versa. The enlightened twenty-first-century mind does not choose between science and spirituality; it marries the two. And there is no better guide for how to do that than Ervin Laszlo.

In Laszlo's words: *To find our way toward a better life and a better world we need science as well as spirituality. Science without spirituality misses the intuitive elements of human experience; elements that many great scientists have valued on a par with, and even above, reason and logic. But spirituality without science cannot offer reliable guidance for confronting the problems we face in the world. We need both science and spirituality, and we need them together, coherently linked. The need is for a dedicated and lasting alliance between science and spirituality.*

While most of us have a pretty good understanding of what science means, the word *spirituality* seems fuzzy to many. What makes this book so intriguing, and relevant, is that it presents spiritual laws in the context they deserve: as discernible, unalterable laws that prevail within our internal experience as fundamental as the scientific laws that prevail within our external world. Though invisible to the physical eye, what Laszlo refers to as the Akashic Field is no less a fundamental factor in human experience than is gravity or the weather.

In addition—and incredibly—not only the human spirit but even the human brain is coded in ways we had not previously realized to mine our spiritual secrets. Both science *and* spirituality are only beginning to reveal their greatest secrets to the twenty-first-century mind. In the words of Albert Einstein, "The more I know about physics, the more I want to know about metaphysics." Reading this book, you will understand why.

The coronavirus pandemic will cast a specter over life in the twenty-first century, long past the time when the current crisis ends. It will leave in its wake not only a greater search for scientific answers but also a greater search for meaning. And as is eternally true, humanity will mysteriously have been blessed by opportunities made available by even the most terrible thing. In the deepest parts of the night lie the invitation to morning.

Read this book and you will understand why.

INTRODUCTION

Our Quest for a Better Life and a Better World—in History and Today

The wish to understand and to improve the world are perennial aspirations of the human psyche. The search for this understanding and improvement has a long history.

In times past, our fathers and forefathers conducted their search under the influence of the religious and spiritual doctrines they inherited from their own predecessors. But today, the problems of everyday life have overshadowed this search. Except for deeply religious and spiritual persons, the probability of gaining enlightenment from the wisdom evolved by previous generations has been seen as insignificant. But today, as we pull out of the virus pandemic and seek the way forward, this is no longer the case. The search can be, must be, and is in fact, revived.

The quest in modern-day thought

In the history of modern-day thought, more and more people have freed themselves from the constraints of preconceived doctrines. This process has reached a new milestone in our day. Now we are looking around, and asking ourselves: who are we? And what kind of a world is it that we live in? Before we enter into the substance of this book, we should look back for a moment, and ask about the history of our answers to these perennial questions.

* * *

In the early Modern Age, our quest was inspired by the power of our imaginations. We found ourselves in a universe replete with meaning but also with mystery. This was the era of eighteenth- and nineteenth-century romanticism and metaphysics. We lived in a universe watched over by higher powers and intelligences, both benevolent (divinities and angels) and malevolent (evil spirits and devils). We attached meaning to everything we saw and everything that happened to us and those around us.

We soon realized that we cannot grasp everything in this meaningful but deeply mysterious world. We recognized that there are things in the world that surpass our understanding. Some of these things are beyond the scope of the world we can perceive with our bodily senses because they are too small or too large, or too far from us. But there are also things that are beyond human understanding. We did not wish to attribute these things to the inscrutable mind of God; we preferred to place their source in a deeper dimension of reality. We envisaged a cosmic source beyond rational understanding. This tenet inspired many great philosophical schemes, and a plethora of masterworks in literature.

The order of nature turned out to be a great and mysterious reality, and so are the vagaries of human destiny.

Who are we in this world—and why are we faring the way we do? Is it karma, or just accident? These questions did not have a clear answer. Nonetheless, we pursued our search.

Then came a time for caution, for second thoughts. With the advance of science, we witnessed the birth of the scientific world picture, and that picture was less mysterious, but also more disappointing than the fruits of unbounded imagination. We had to choose between what we instinctively and intuitively believed was the real world, and the purified world of the theories and findings of modern science. Many of us opted for going with the tenets of science. This called for weeding from our established world picture the speculative elements, and affirming only those that science could warrant as an aspect of the real world.

A new phase opened in our quest for living a better life and a better world, an age dominated by down-to-earth thinkers: the positivist philosophers. They told us to get the facts, and only the facts. Under "facts" they meant data we could measure, repeat, and record. Find what the data tell you is the real world, and ask neither why what you find is there, nor what else could also be there.

The fact-reduced world of positivism proved to be a grounded but sterile world. The "facts" we encountered did not allow further questioning and proved to make for a meaningless world. Positivists themselves said that the real world is "just one darn thing after another."

Our search entered a further phase. We realized that our data-reduced world is sound, but not adequate—it is devoid of meaning and likely only pierces the surface of reality in the world. The real world lurks somewhere in the background of the data, but little of it can be known by the purified data-reduced world of positivist science.

Some of us ventured beyond the limits of positivist science and found ourselves in a strange universe. It contained things and events that were inexplicable in light of common sense, and even in terms of the concurrent scientific worldview. Science moved beyond the commonsense world, but society as a whole did not. The mainstream's world picture remained dominated by the Newtonian concept, where all things are ensembles and combinations of material particles, obeying unchanging mechanistic laws of motion. There are no exceptions to this state of affairs, and no grounds for denying its truth.

This turned out to be insufficient to explain what science was beginning to tell us are the contours of the real world. Scientists have come across experiences that appear as physical things and events, but are not confined to specific points in space and time. They have found phenomena of consciousness that are not limited to the brain. These findings were anomalous for the dominant world picture, yet they cropped up insistently in the wake of the ever wider and deeper investigations undertaken at the leading edge of science. We could no longer either ignore or deny these paradoxical findings.

We had to resolve the conflict between science's findings and what we could believe is the real world. The difficulty in accepting science's account of it was compounded by the fact that more and more of the world described by scientists shifted beyond the scope of human experience. It turned out that Shakespeare was right: there is more to this world than you and I can even imagine.

Where we are now

How can we account for the extra- or nonsensory elements of our experience of the world? We cannot go back to the wonderworld of speculative philosophy—that era has now been left behind. We have to recognize that the scope of the world discovered in science is wider and deeper than we thought. It includes elements that are inexplicable both for the classical view of the world, and for the emerging view in science.

Throughout the twentieth century, science's world picture kept expanding; it embraced more and more in space and in time, as well as in depth and complexity. This expansion remains in full swing in the twenty-first century. Positivism performed a much-needed service, a thorough housecleaning. It swept away the unprovable assumptions and speculative theories of the previous era. But the purified world it bequeathed us was full of gaps. It could not account for the phenomenon of consciousness, and for things and events that exceeded the known bounds of space and time. It forced us to admit that some elements and aspects of the world are not rooted in experienced "facts." In the past, we regarded these elements and aspects as esoteric, spiritual—in other words, unreal. We excluded them from our concept of the real world. But by the dawn of the twenty-first century more and more evidence has accumulated that suggest that some of these aspects and elements belong to the real world. We need to accommodate them in our world concept.

The gap between the world concept suggested by the investigations of scientists and by our own experience of the world has shifted from the terrain of scientific-philosophical theorizing to questioning our most basic concept of who we are, and what is the world around us. Science renounced the positivistic reduction of theories to facts, but it did not produce a comprehensible world picture. The injunction of Niels Bohr's Copenhagen school of quantum mechanics—that science is not there to solve problems of meaning, and so if you work in quantum physics, "just solve the equations and don't ask what they mean"—did not help. It merely sweeps the problem under the rug.

We had to admit that in understanding the world, the new science, first and foremost quantum physics, substituted one set of puzzles for another. Without an intelligible framework, phenomena such as wave-particle duality, time-and-distance-independent nonlocality, and a host of related findings are just "phenomena," floating in the limbo of an implied but profoundly mysterious world. Quantum physics proved to be even more mind-boggling than the speculative metaphysics of the eighteenth and nineteenth centuries.

Where do we go from here?

It is time to move on. We should no longer proceed on the assumption that the findings of science offer a full and final answer to our query regarding a better life and a better world. We need to go beyond science, but not fall prey to unfounded speculation. The answers we seek could be harbored in the millennia-old accumulation of wisdom in the spiritual traditions. Can this wisdom be injected into the insights offered by science? This could be important because there may be more to the accumulated wisdom than illusion and fantasy. If there is indeed more, our search for a better life and a better world would be enriched by a new and profound resource.

There are many spiritual insights to which contemporary science could fruitfully be connected. To cite but the most immediately relevant, there are insights such as the intrinsic interconnection among all things, the conservation of the trace of all things, and a consistent trajectory in the developmental path of all things that are common to both science and spirituality. These insights are present in the body of inherited wisdom the ancients called the Akasha. The Akasha is not a random sphere of being, but a highly ordered and consistently evolving sphere. It is governed by immutable laws. These laws can guide our search for a better life and a better world. This book is dedicated to facilitating our access to this guidance by examining these laws, and understanding how they can help us to live better, in a better world.

PART 1

THE LAWS OF THE AKASHIC FIELD

THE CONNECTIVITY LAW

With remarkable insight, five thousand or more years ago the Hindu seers spoke of a dimension of the world that would be prior to, and more basic than, the others. The Akasha dimension, they said, is the fundamental dimension of reality—it is more basic than the dimensions of *vata* (air), *agni* (fire), *ap* (water), and *prithyi* (earth). The Akasha harbors the immutable laws and regularities that govern existence in the world.

The laws recognized by the Hindu seers as governing existence in the world include the law that accounts for the interconnection of all things with all other things, and the law that accounts for the conservation of the interconnected things. Together, these laws define a world that is seamless and whole, with all its elements connected in space and conserved in time.

These laws apply to the universe we inhabit, but they do not account for all the processes that take place in this universe. Specifically, they do not take account of the coherence and consistency of fundamental change and development in the world. Unlike some Hellenic thinkers, such as Heraclitus, for whom change was the predominant feature of existence, coherent and consistent change was not recognized by the ancient seers. They held, as Parmenides did later, that existence is fundamentally *being; becoming* is a secondary and possibly illusory phenomenon.

However, today we know that in the universe change is fundamental, coherent and consistent, and real. The world not only *is,* it also *becomes.* This was recognized by the renowned yogi Swami Vivekananda in the early twentieth century. He described the Akasha as the ground of a cyclic process of continual change: evolution followed by devolution, and then by renewed evolution. The Akasha, he said, *becomes* all things in the universe, and at the end of the phase of becoming, all things return again to the Akasha.[1]

Following in the footsteps of Vivekananda, we can complement the two principal laws of the Akasha with a third law: the law of change. This law, the same as the others, does not apply to itself. Unchanging, immutable laws govern change in the world. Thus there are three immutable Laws of the Akasha dimension: the time-honored Connectivity Law; the equally ancient Memory Law; and the third law, which takes account of coherent and consistent fundamental change: the Coherence Law. We now describe each of these laws in turn.

UNIVERSAL INTERCONNECTION IN SPACE AND TIME

Connection among the things and events in the world is a perennial intuition. In contemporary science, it is borne out by the principles of quantum physics. There are no entirely disconnected things and events in the universe. All things are in some way and to some degree interconnected. Although the Hindu seers did not call it that, "nonlocal" interconnection is a basic feature of the true concept of the universe.

Until quite recently, nonlocality was metaphysics and not science. But with the advance of cutting-edge physics, this has changed. For today's quantum sciences, the nonlocality of events is a valid proposition. Quantum physicist Henry Stapp called nonlocality the most profound discovery in all of science, and French philosopher Pierre Lévy pointed out that "the recognition that our universe is nonlocal has more potential to transform our conceptions of the 'way things are,' including who we are, than any previous discovery in the history of science."[2] The universe, it appears, is a *nonlocally whole macroscopic quantum system*.

Matching the new and the old world concept

The world concept of the quantum sciences is radically different from the mainstream world concept of the Modern Age. The world is not an arena for the motion of solid, indivisible particles in passive space and indifferently flowing time. This is not a world of separate and separable parts, where things occupy single positions in space and time. It is not a complex mechanism fashioned of matter. If we were to define it with a single concept, we would say that the world is a *hologram*.

In a hologram, as we know, all the information that

constitutes the image is present at all points. That which is here is also there; it is basically everywhere.

The world as a hologram is even more fabulous than this. It is not only beyond limitation in space; it is also beyond limitation in time. What is here today was here yesterday and will be here tomorrow. All the information that codes the system is present in each of its parts, and this information does not vanish. The information that makes the universe what it is, is both spatially and temporally nonlocal. And it is present in every particle and in every atom—and in every organism, including you and me.

This seems like a fabulous idea, but there is solid reasoning behind it. The world is not "material"—it does not consist of anything we could call matter. Physicists have not found anything in space and time that would correspond to the idea of a material substance. What research on the universe discloses is information and energy. The entities of the physical world are clusters and configurations of *informed energy*.

The world "runs" on information, but on what it runs is not the garden variety of information we think of when we read a newspaper or talk to a neighbor. The world runs on "in-formation." It runs information that is correctly spelled with a hyphen. [3]

The concept of in-formation comes from quantum physicist David Bohm. In-formation is an active impulse that acts on, "in-forms" things and events, but does not convey energy in the usual form. It is not the kind of energy that can run down and dissipate. It is a "nonvectoral" dynamic impulse at the core of the universe. It forms all things; in fact, it forms the universe. It is what makes the universe what it truly is, a sphere of coherently structured configuration of in-formed energy, and not a random swirl of inert gases.

In-formation corresponds to the classical notion of a vital impulse, or perhaps divine law or cosmic consciousness, that would shape the observed world. Today in-formation is a new and essential tenet of physical science.

In-formation accounts for the presence of the particles that make up the basic entities of space and time. These entities are "in-formed" by an impetus that is itself beyond observable space and time—beyond the manifest domain of the universe. This extraordinary impetus brings and holds together the nucleons that constitute the core of the atom, and brings and holds together atoms in molecules, molecules in crystals and macromolecules, and these again in cells and organisms. Max Planck was so impressed with this impetus that he said that we must assume the presence of a higher intelligence in the universe to explain it.[4]

The world is a system of systems of in-formed energy, constituted and maintained by a nonenergetic cosmic impulse: "in-formation." Some systems of in-formed energy are sufficiently stable and coherent to appear as material things, but to assume that this is what they are, as Einstein pointed out, is an illusion.

In addition to positing the presence of a nonenergetic impulse in the world, and negating the existence of matter, a further revolutionary insight is shaping the world concept of contemporary science. *The universe we observe is not all there is in the world.* This is a revolutionary tenet as well, because the idea that there would be a world beyond the observed universe is strange and unacceptable for mainstream science. For the past 250 years, scientists maintained that everything that exists is material, and all material things are part of the universe. Now it appears that there are no material things in the universe at all—there is no such thing as "matter."

We know that the universe we observe and inhabit is not all there is in the world. The universe we inhabit is not *the* universe; it is only *our* universe—just as *this* solar system and *this* galaxy are just "our" solar system and "our" galaxy. There may be millions or billions of other solar systems and galaxies in the larger context of the cosmos, and perhaps many universes as well.

The current understanding is that the world did not come into being when *our* universe did: there was a prespace in the world before our universe was born, and there may be a postspace as well, persisting when our universe has left the stage.

According to the currently accepted cosmology, the universe we inhabit was born in the aftermath of the cosmic explosion known as the Big Bang about 13.8 billion years ago. It will vanish when the processes that make up its quasi-material furnishings have run their course. Stars and galaxies will either become part of an eternally expanding "dead" universe, remain balanced at the razor's edge between expansion and contraction, or become part of a universe collapsed to quantum dimensions—to a dimension smaller than the head of a pin.

Regardless of the ultimate destiny of *our* universe, it is clear that it emerged in the embrace of a wider, perhaps infinite and eternal world, the world the Greek philosophers called the *Kosmos*. Our universe is a local and temporary domain of the Kosmos.

Interconnection: the "meta" physics

We turn now from the exploration of the old and the new world concept—or "metaphysics"—to the specific features of the presently validated world concept. One such feature is interconnection. The tenet of interconnection tells us that throughout the universe, things and events are intrinsically interconnected.

In the quantum sciences, this tenet is not metaphysics but physics. Universal interconnection has a revealed physical basis. This exceeds the scope of classical physics: that body of thought cannot explain it. For an explanation, we need to turn to the theory of "in-formation" proposed by Bohm. In Bohm's concept, there is the manifest universe (the explicate order), and there is also the nonmanifest but equally or perhaps even more real "implicate" order. The perceived explicate order is in-formed by the underlying implicate order.

In simple language, the implicate order "in-forms" (shapes and structures, creates and orients) the explicate order. The explicate order is in principle observable, and the implicate order is not. Its existence can only be inferred from the observation of the entities and processes that appear in the explicate order.

Now we apply this world concept to our understanding of the world.

Quantum researchers find that there are no categorical breaks and separations among the things that furnish the world. These things (considered "events") constitute a seamless whole: a continuum. It is best conceptualized as a *field*. The concept of a field is the simplest and most logical way to map the continuum disclosed in the quantum sciences.

"Field" is now a basic element in science's understanding of the world. It has become basic since the publication of the so-called Yang-Mills theory in 1954.[5] In this "gauge" theory of physical reality, the elementary particles of the universe are not tiny billiard balls. They are intrinsically interconnected excitations in a universal field.

The concept of a universal field is consistent with the vision of the ancient seers. What they called Akasha is a dimension or order of the cosmos with field-like characteristics. We distinguish three characteristics that match the traditional Akashic concept with the current understanding: universal connectivity (discussed above), systemwide memory (to be discussed below), and continuous coevolution (to be discussed in the next section).

Universal interconnection in the field

Avant-garde quantum science makes clear that the objects we observe are not what they appear: they are not separate entities but excitations in a universal field. They are not separate objects, and they do not even exist unless and until we observe them. With a slight overstatement (often committed even by quantum scientists) we can say that their existence is due to our observing them. The things we observe are embedded in a sea of potentiality; they are "fished out" from this sea by our observation. Our observation "actualizes" them.

It is sometimes claimed that the observation—and hence the observer—"creates" what he or she observes. This is the somewhat exaggerated creation doctrine of quantum physics. (The phenomenon is more correctly termed the observer effect.) It encourages countless speculations. Some researchers (Martin Rees among them) claimed that the universe itself came into existence because it was observed. It does not matter that the observation took place billions of years after the birth of *this* universe. The principle maintains that the universe exists, and has existed all this time, because we now observe it.[6]

In a more technical vein, what quantum physics tells us is that the observation of an object collapses its wave function. This collapse means that the object's wave function (which consists of the superposition of all its possible states) is reduced to a single so-called *eigenstate*. Then we can describe the object as a classical entity with determinate properties and a unique location in space and time.

The observation-induced collapse of the wave function appears to "create" the object—it lifts it from potentiality into actuality. Here the idea of "creation" is an exaggeration and is not to be taken literally. It is not as if the observer would create the object ex nihilo, out of nothing. The observer merely lifts the object from the level of potentiality into that of actuality.

Let us consider now how this seemingly abstract theorizing relates to interconnectivity in the real world. Quantum theory holds that the objects our observation calls up from the level of potentiality into actuality are intrinsically interconnected. They are not separate objects linked by some form of connection or energy, but seemingly separate excitations in a continuous field. They are interconnected since they are embedded in the field. This seemingly abstract postulate has time and time again been tested in controlled experiments by physical scientists. Nonlocal connections have come to light between quantum particles even when they are separated by considerable distances. It turns out that every particle that has ever occupied the same quantum state as another particle remains instantly and enduringly connected with it: their quantum state remains correlated.

Nonlocal correlations are not limited to the microdomain of the quantum. Nonlocal connections have been discovered within living organisms, as well as between living organisms and their milieu. The universe as a whole manifests fine-tuned spatial and temporal connections that correlate the state of the entities that exist in it. Physicist Erwin Schrödinger called the persistent correlation between the state of distinct and even distant particles "entanglement." The expression enjoys wide currency. We can conclude that the quantum world is an entangled domain, with instant interconnection in the form of nonlocal correlation between all its elements.

Nonlocality, considered as intrinsic universal connectivity, is inexplicable for the Newtonian world concept; it defies its limits of space and time. However, the mystery of nonlocal connectivity is mitigated, if not entirely eliminated, when we realize that the Akashic Field of nonlocal connectivity is actually a hologram. The hologram offers the simplest and the most logical explanation of nonlocality. A hologram contains all the information that creates the image in all its parts. Hence the parts are holographic projections, and not separate entities. The connection between them is not the connection of one separate entity to another, but the connection of one part of a set of projections to another part. Separation between the projections is illusory: after all, they are parts of the same hologram.

The illusory nature of separation between things was already recognized by Plato and illustrated in his famous Allegory of the Cave. Here prisoners are chained to a cave in a position where they cannot turn around; they can only see the wall in front of them. They see shadow figures on the wall that move and interact, and they mistake them for separate individuals. In reality they are shadows cast by one and the same light source—a fire burning behind them.

The surface on which the holographic images we take for separate objects are projected is the explicate order: the manifest dimension of the universe. The projections that appear to us as separate things and events are projections of the cosmic hologram which is the implicate order, on the explicate order, which is our manifest reality.

The universe, the totality of the reality we observe, is an embracing information-and-memory field, governed by distinct, and as far as we know, immutable, laws. These laws include the three laws that define that aspect of the universe we call the Akashic Field.

THE MEMORY LAW

We now turn to the second immutable law of the Akashic Field: the Memory Law. This law accounts for the conservation and recall of the content of our experience—the content of all human (and presumably also nonhuman) experience.

This ancient tenet is likewise borne out by the findings of contemporary science. In the worldview of the quantum sciences, the entities to which our experience refers are not material things and events, but by *information*. It is information that structures the vibrations we experience as material things and events. The Memory Law accounts for the conservation of this information, and for its recall. In the language of the information sciences, we can say that everything that happens, everything that emerges and has ever emerged in space and time, is "saved"—it continues to be present in the Akashic Field. Information in the universe cannot be canceled or obliterated; it can only be transformed. And it can also be "saved."

This insight is an integral element of the wisdom traditions. Everything that happens in the world, the yogis maintained, is vibration, but not everything is vibration on one and the same plane. The things we experience are vibrations on various planes, and when they shift from one plane to another, we may no longer be able to perceive them. But they do not disappear—they continue to exist on a different plane.

The Swami Vivekananda affirmed this concept. He said that the universe is an ocean of ether, made up of layer after layer of vibration of different kinds, on different planes. It is evident, he said, that those who live on the plane of a certain vibration have the power to recognize one another, but they do not recognize those who live on a plane above or below them. Yet we are all vibrations of the same world, differing only in the frequency of our vibration. The goal of spiritual masters, according to Vivekananda, is to reach samadhi, the state of "superconsciousness" where they recognize beings on other, and perhaps all, planes of vibration.

Vivekananda claimed that the universe is a unitary domain, composed of two materials, one of which is perceptible as matter, and the other is a more etheric substance called the Akasha. The Akasha is an omnipresent, all-penetrating dimension of the cosmos. Everything that has form, everything that is the result of combination, said Vivekananda, evolved out of Akasha. The Akasha became air, liquids, solids; it became the sun, the earth, the moon, the stars, the comets. It became the human body, the animal body, the plants, every form that we see, and everything that exists. At the beginning of creation there was only the Akasha. And at the end of the cycle of existence the solids, the liquids, and the gases melt into the Akasha, and in the next phase of creation they arise again from that deep dimension.

This view is perfectly in accord with contemporary quantum science's concept of the world. In the quantum concept as well, the universe is a field of vibration. When the vibrations of which things are composed are in phase, they produce the phenomena we know as the things and events of the world. Beyond the appearance of things, there is a deeper dimension of reality. The clusters of in-phase vibration constitute standing or propagating waves originating in that dimension, and they manifest in the observable world at various domains of frequency. Our senses only register vibrations within a given, and relatively narrow, frequency range. That is the range we need to apprehend in order to live, and to survive as a biological being. We are clusters of waveform vibration, subsisting in the frequency domain that permits sensory perception. We vibrate on the same "plane" as other biological beings on the planet and thus we can perceive them and interact with them. This plane constitutes our vital environment.

According to the cutting edge of contemporary science, clusters of vibration constitute everything that exists in the manifest universe, beginning with the quarks, the quanta, and the "spherical Planck space-time units" that compose atoms and may be the smallest discernible units of informed energy in the world. But the clusters of vibration we perceive are not *all* the clusters that exist in the world: we perceive only those we need to ensure our existence.

Thanks to modern science and the technologies it has spawned, the relatively narrow range of our perceived reality is enlarged. Space-based telescopes and electron microscopes are shining examples of our power to extend our perception beyond the spatial range of our bodily senses. At the super-small scale we apprehend quarks and space-time units, and at the superlarge scale we come across black holes, gravity waves, and the birth and death of stars and galaxies. All items in this technologically enabled domain of perception refer to vibrating clusters of interfering waves. We do not encounter extra- or superphysical realities, only clusters and superclusters, or segments and segments of segments, of the structured vibrations that exist in space and in time.

Considered in its totality, the universe is a super-supercluster of waveform vibration. The memory law of the Akashic Field tells us that this super-supercluster (and every part and element of it) is "saved."

THE CONSERVATION AND RECALL OF EXPERIENCE

We have said that the presence of clusters of waveform vibration suggests that there is real and active information in the universe, and that this information is not evanescent. The evidence for this affirmation becomes clear when we consider its contrary. Failing the conservation of the information that underlies the structures and processes of the universe, the universe would be a random heap of unrelated events. The universe, however, is a highly ordered ensemble of interconnected events. A universal information-and-memory field conserves the information that creates and maintains the universe. Nothing that has ever taken place in space and time fails to enter the universal information field and be conserved in it.

This basic hypothesis applies to all aspects of human experience. Here, however, we shall concentrate only on its relevance to our own life: the conservation and recall of our own experience.

There is empirical evidence supporting the assertion that the totality of human experience is conserved. The evidence is that our experiences are lasting "memories"; they are not subject to erosion and attrition.

Our experiences are permanent memories present in the deep dimension of the universe, and in principle they are always available for conscious recall. Experimental psychologists find that under suitable conditions people can recall practically all experiences they have ever had—for example, in so-called spiritual emergencies, in altered states of consciousness, and under hypnosis.

A seemingly esoteric but well-documented variant of this recall is known as the life review. It comes about when the person experiences a deep trauma and appears to be close to the portals of death. This is a recall and review of everything the person has experienced in his or her lifetime—and it occurs independently of a functioning brain. This is the phenomena known as near-death experiences (NDEs). People in the near-death condition seem capable of recalling the full range of their lifetime experiences, and their recall can occur during the time their brain functions are flatlined. This puzzling recall has been subjected to serious research and analysis, reported among others by Pim van Lommel.[7] The lifetime recall does not occur in every NDE, but it occurs in a significant number of cases—in some surveys, up to 48 percent of the people surveyed have reported it.

Thus there is significant evidence that the experiences of individuals do not vanish without trace: they can be recalled, at least in part. The question is, recalled from where? Where are our and all human experiences stored for recall? Clearly, they are not stored in the brain, since they can be recalled in the absence of brain function. It is likely, then, that they are stored extra-somatically, in the information field revealed in quantum physics and quantum cosmology and identified here as the Akashic Field.

It is logical to consider that our experiences enter the Akashic Field and are conserved in that field. The Akashic Field is a "live" storage medium from which selected items can be recalled. This has been anticipated in the ancient and perennially recurring intuition of the Akashic Records. The Records manifest experiences had by people at any and all points of their life. [8]

We do not reexperience our lifetime experiences in the ordinary state of brain and consciousness. These experiences are stored in a band of frequency other than that of ordinary experiences. The extra-somatic information-store functions at nonordinary frequencies. These are accessible when the normal scope of our brain functions is extended. Then highly intuitive persons "read" the Akashic Records. Their brain is then tuned to the frequencies of the Records.

We know that in normal states of functioning, the brain decodes only a limited range of the spectrum of the frequencies that are available to it. We know this experimentally in the case of the electromagnetic field, where the visible spectrum of the EM field is only a tiny fraction of the total spectrum of vibrations. Normally only a narrow band of the EM field's vibrations are processed by the human brain, but in altered states a wider range of frequencies becomes accessible. Yogis and spiritual masters decode a wider-than-ordinary range of frequencies. Prophets and visionaries are not necessarily delusionary dreamers: they may be accessing a wider band of frequencies in the world than people in ordinary states of consciousness.

In near-death and other altered states of consciousness, frequencies are accessed that are beyond those accessed in ordinary states. All frequencies are conserved in the Akashic Field and are recallable from that field in the corresponding states of consciousness.

THE COHERENCE LAW

We now turn to the third of the immutable laws of the Akashic Field, the law that accounts for the emergence of order and structure in the universe.

As noted, the contemporary sciences, especially the branches known as the quantum sciences, tell us that the universe is not a collection of separate entities and events. It is an integral quantum system governed by immutable laws. The first law of the Akashic quantum universe is universal interconnection among all things, and the second law is the conservation of the information governing the interconnected things. Now we add the third law: the law that accounts for the fact that the systems that emerge and evolve in the universe are not random aggregates of their elements, but structured integrations of them. This is the Law of Coherence.

Before considering the evidence for this law, we should define what we mean by "coherence." In the context of science, we need to distinguish two forms of coherence:

quantum coherence, appearing on the ultrasmall sub-quantum scale, and macroscopic coherence, applying to systems at the meso- or macroscale.

Quantum coherence is the superposition of the wave function of microscale quanta in a way that they are no longer distinguishable as individual entities: for all intents and purposes they are one. *Macroscopic coherence,* in turn, is the tuning of the parts or elements of a system in a way that ensures the sustained functioning of the system in its environment. When so tuned, the diverse elements of the system behave as one and respond to the world around them as one. Macroscopically coherent systems are individually distinguishable, but they are not categorically separate from one another.

Quantum coherence has been discovered in the famous experiments where a quantum particle (more exactly, the quantum state of a particle) is divided, and one-half is projected to a finite distance. It is found that over any hitherto found distance, the states of the particles remain correlated, as if they were physically one. Macroscopic coherence, in turn, appears in complex systems where the parts or elements of the systems are tuned to receive, and respond to, all other parts or elements as one.

The contrary of quantum coherence is "decoherence." Decoherence is the state of objects in which they appear as classical entities, possessing their own definable characteristics, and unique position in space and time.

Here we are concerned with the state of macroscopic coherence in the manifest state of systems. This is the state in which we, as other living systems, find ourselves.

Evolution creates coherent systems in space and time. These systems are both internally coherent, and coherently related to other systems in their surroundings.

We now turn to the question of evolution in the universe. A common feature of cutting-edge theories is the recognition of a nonrandom directionality manifested in the evolution of the emerging entities. Theories addressing the cosmological evolution of stellar systems and galaxies agree with theories concerned with the evolution of biological systems in recognizing an overall directionality of the evolutionary process. In whatever form and at whatever scale of size and complexity the process unfolds, it tends to maximize the level of coherence of the evolving systems.

EVOLUTION TOWARD COHERENCE

The universe being nonrandom, the behavior of systems in it follows basic laws. The Law of Coherence governs the direction in which the systems evolve in space and time. At all scales of size and complexity, this direction is marked by a progressive, but not necessarily smooth and linear, increase in the systems' level of coherence. This is a general statement referring to universal processes, and it is supported by empirical evidence.

A look at the evidence

Cosmologists discovered an entire array of curious "coincidences" among the physical parameters of the universe. For example, the mass of elementary particles, the number of particles, and the forces between them display harmonic ratios. Many of the ratios among basic parameters can be interpreted on the one hand in reference to the relationship between the mass of elementary particles and the number of nucleons (particles of the atomic nucleus) in the universe, and on the other in reference to the relationship between the gravitational constant (the factor of gravitation in the evolution of the universe), the charge of the electron, Planck's constant (a unit of measurement used to calculate the smallest measurable time interval and physical distance), and the speed of light.

The coherence of the basic parameters of the universe is not a mere coincidence: it is a fundamental precondition of the emergence and evolution of life. Complex systems can only evolve in a universe of which the physical constants are precisely correlated. Variation of the order of one-billionth of the value of some of these constants (such as the mass of some quanta, the speed of light, the rate of the expansion of galaxies, and two dozen others) would have resulted in a sterile, lifeless universe. Even a minute variation would prevent the creation of stable atoms and stable relations among them and would thus have precluded the evolution of systems that manifest the phenomena of life.

The evolution of the complex and coherent systems is not just a serendipitous happening reserved for a few fortunate types of systems in a few fortunate regions of the universe. The universe appears to be a template for the evolution of life: this evolution takes place in many places and occurs in many forms. The forms of life that emerge in the universe range in size and complexity from atoms and molecules to cells and organisms, and from individual organisms to societies and ecologies of organisms, and entire biospheres.

It is astronomically improbable that cells and organisms, and societies and ecologies of organisms, would have come about through a random interaction among their components. The statistical analysis of even relatively simple systems reveals that to produce them by a chance shuffling of their components would have taken longer than the age of the universe. To use a simile suggested by mathematical physicist Fred Hoyle, the probability that complex systems such as living organisms would have come about through a random mixing of their elements is like the probability that a hurricane blowing through a scrapyard would produce a working airplane.

When it comes to biological systems, the probability that they would have been produced by random processes is further reduced. The complexity of the DNA-mRNA-tRNA-rRNA transcription and translation system precludes that such systems would have been produced by random processes. While it is true that in an extended process almost anything that could happen will happen, the length of time required for the evolution of biological systems by random interactions exceeds all reasonable finite and hence all reasonable probabilities. Some 13.8 billion years for the evolution of particles and atoms and other physical entities, and about 3.8 billion years since the appearance of living systems in the universe, are not sufficient to account for the presence of stars and galaxies and of the web of life on this planet.

The evidence for the universal evolution of complex

systems, including those that exhibit the phenomena of life, is cogent and convincing. It is important to recognize and take into account, because it disproves and replaces a basic credo of mainstream science. Mainstream scientists since the time of Newton have been saying that chance interactions produce the phenomena we find in space and time. The contrary assertion, they feared—namely that systems evolve in accordance with a preestablished plan or design—would raise the specter of teleology: it suggests the presence of a higher will or mind. Scientists preferred to stay with the theory that the phenomena we encounter are fully accounted for as resulting from an undirected succession of states in which each state sets the conditions for the next.

The challenge to this dogma of mainstream science is raised by the complexity of the phenomena we find in the universe, coupled with a finite time for their emergence. Statistical analysis shows that the phenomena we encounter are too complex to have been produced by chance interactions within the available time. The maximum available time is about 13.8 billion years since the beginning of the system-building process in the aftermath of the Big Bang. This time span, although enormous, is insufficient to account for the complex and coherent systems that emerged in the universe. "Something" other than chance must have been governing the interactions that produce the systems we observe.

Notwithstanding the fears of mainstream scientists regarding teleology, we need to contemplate the possibility that an intrinsic impetus in the universe is orienting the processes of evolution in a nonrandom direction.

A capsule history of systems evolution

The particles that formed following the Big Bang began to assume form and structure when the universe cooled sufficiently for stable or quasi-stable entities to form. These entities are quarks, as well as quantum particles such as leptons (electrons, muons, tau particles, and neutrinos), mesons (pions), and hadrons (baryons including protons and neutrons). Quarks form the particles, the protons and neutrons and other short-lived entities that compose the nucleus of stable atoms; they are bound together by gluons in the nuclear field. Protons themselves are composed of quarks: two "up" quarks and one "down" quark. At suitable temperatures, the protons bind with electrons, capturing them in the "electron cloud" of atoms.

Over time, hadrons such as protons built from combinations of quarks to form the nucleus of the various atoms. The number of protons in the nucleus is the defining property of the atoms that emerge. The resulting atomic structures fill the periodic table of the elements. They encompass atoms from hydrogen to uranium, and beyond.

The atoms that emerge in the universe build into more and more complex entities. They build into molecules and multimolecular assemblies on the surface of this planet, while on the astronomical level they create stars and stellar systems and entire galaxies.

As already mentioned, the processes of evolution do not take place in flat euclidean space and in independently flowing time. They occur in a highly structured field that sets the probabilities of building complex systems out of combinations of simpler structures. This means that evolution unfolds in a structured medium, allowing complex stable or semi-stable systems to emerge as integrations of simpler entities.

Although physical systems are the prior template for the evolution of biological systems, evolution was recognized in the biological sciences before it was recognized in the life sciences. The realization that the evolutionary processes occur also in the physical domain dawned only in the first decades of the twentieth century, when Einstein's hopes for an eternally unchanging matrix-universe proved unrealizable. The universe of general relativity proved to be undergoing time-dependent evolution.

The nonrandom process we call evolution builds toward increasingly complex and coherent systems. "Something" orients this process in an identifiable direction, but there is no consensus on what that something would be. Henri Bergson speculated that it is an *élan vital* that counters the trend toward the degradation of energy in natural systems; biologist Hans Driesch suggested that it is a counter-entropic drive he termed entelechy. Philosophers Pierre Teilhard de Chardin and Erich Jantsch postulated a dynamic tendency they called syntony, while others called the universal evolutionary drive syntropy. Eastern thinkers called it prana, a Sanskrit term, while in the West Wilhelm Reich suggested that the driving force is the energy he called orgone. In turn, Rudolf Steiner said that evolution is an etheric force, and Newton himself sought to incorporate such a universal force or "spirit" into his theory. The mechanistic laws, he said, are not a full description of reality; they need to be completed with the recognition of an "enlivening and ensouling" spirit in all things—a "spirit of vegetation."[9]

The nature and origin of the universal "spirit of vegetation" was not definitely known. It remains mysterious to this day. Scientists sometimes resort to supernatural explanations. Examples are Planck's affirmation that a higher intelligence is behind the force that holds particles together in atoms, and Einstein's conclusion that anyone who has studied the laws of nature must conclude that underlying the laws is a mind infinitely higher than our own.

Failing natural explanations, resorting to extra- or supernatural forces is understandable, but it is a last resort. Science is based on explanation in reference to natural processes. Such an explanation was not available until recently. Now it is available, thanks to Bohm's theory of "in-formation." As already noted, the observed world, the explicate order, is an in-formed domain of space-time. Its in-formation exhibits an impetus that we assume has its roots in the deep dimension of the cosmos, the implicate order. It is expressed in the form of immutable laws. These laws in-form the explicate order and account for the nature and the evolution of the order we find in it.

To conclude. The connectivity we find in the universe is the effect of the *Connectivity Law* of the Akashic Field; the conservation and potential recall of experienced events is the effect of the *Memory Law;* and the nonrandom evolution of systems is the effect we can attribute to the *Coherence Law.* Together, these immutable laws account for the emergence and presence of order and structure throughout the universe.[10]

THE VIEW FROM SCIENCE (I)

What Is the Akashic Field?

What is the Akashic Field—is it fact or fancy? Does science admit the existence of this field, and does it offer valid grounds for describing it? The new physics, in particular quantum physics, offers an answer. The answer is yes: although we may call it by other names, the Akashic Field exists. It is a universal information-and-memory field—it is how the universe appears to an observed within it.

QUANTUM SCIENCE AND INFORMATION IN THE UNIVERSE

The recognition of the existence of the Akashic Field—as the appearance of the universe to us—is supported by a newly discovered and surprising feature of physical reality. Information exists objectively in nature; it is not only something we produce. It exists independently of human minds and human beings.

Information has existed in the cosmos at least since the birth of this universe and may have existed even before then. It is likely that it will continue to exist after not just humans, but this entire universe, will have vanished from the face of reality.

According to *Apollo 13* commander Edgar Mitchell, information is part of the basic substance of the universe. The quantum vacuum, he suggested, is a "holographic information mechanism that records the historical experience of matter." The question for science is, How does this information work? How does it record and transmit the "historical experience of matter"?

We know that interactions between things in the physical world are mediated by energy. Energy can take many forms—kinetic, thermal, gravitational, electric, magnetic, nuclear, actual, or potential—but in all its forms energy conveys an effect from one thing to another, from one place and one time to another place and another time. Energy must, however, be conveyed by something; and if it is conveyed across space, science would consider that it is conveyed by a field. In that case, space is not empty—it is filled with an effectively acting field. Initially, scientists called space in the absence of matter a vacuum, but continuing this use is misleading. Space is not empty. The field that fills it conveys light, gravity, electromagnetism, and energy in all its forms. It also conveys information, that is, the cosmic impetus Bohm called "information."

The idea that information is present in the universe is a recurrent insight in the history of thought. It suggests that information is not an abstract concept: it has a reality of its own. It is an element of the real world. It constitutes a non-material but active and effective continuum: a field.

The evidence for a field that would contain and convey information is not direct; it must be reconstructed in reference to more directly available evidence. Like other fields known to physics, such as the gravitational field, the electromagnetic field, the quantum fields, and the Higgs field, the information field cannot be seen, heard, touched, tasted, or smelled. However, its effects can be perceived. This is the same in regard to the other fields known to science. The effect of the G field is gravitation among separate masses; general relativity and related field theories show that postulating the existence of this field is the simplest and most logical explanation of the observed effects. Now we can add that the effect of the EM field is the transmission of electric and magnetic force, and of the Higgs field it is the presence of mass in particles. In turn, the effect of the weak and strong nuclear fields is attraction and repulsion among particles in proximity. The effect of the information ("in-formation") field is spontaneous correlation and instant interconnection among the particles that constitute the smallest observable entities of the universe.

The information field is a fully warranted addition to the fields known to science. Things and events could not be affecting each other within the bounds of space and time unless they were connected by a physically effective medium: an intervening field. Electric and magnetic phenomena are connected by an electric and magnetic field: the EM field. Michael Faraday's EM field was a local field, associated with given objects. James Clerk Maxwell proposed that the electromagnetic field is not local but universal: it is present everywhere. Modifications of the EM field travel throughout space at the speed of light. A changing electric field produces changes in the magnetic field, and this in turn produces changes in the electric field. The changes are registered throughout the electromagnetic field, that is, throughout the space and time of the manifest universe.

The universal electromagnetic field was a revolutionary insight, for it meant abandoning the notion of empty space as a vehicle for transmitting the forces behind the interaction of particles.

Beyond the quantum vacuum

In contemporary physics, the information field that underlies relations among things is termed quantum vacuum. But, as just discussed, space in the universe is filled and not empty. Therefore the universal field that transmits physical effects is more correctly called quantum (indeed, subquantum) *plenum*.

In the late nineteenth century, most physicists believed that space is filled with the invisible subtle energy they called luminiferous ether. They hypothesized that the luminiferous ether produces friction when bodies move through it, and thus slows their motion. At the turn of the twentieth century, the famous experiments by Michelson and Morley failed to observe the expected effect, and the ether was removed from the scientists' world picture. Einstein's theory of relativity replaced the ether with a four-dimensional space-time continuum. His theory wedded space with time and created a four-dimensional matrix. Repeated experiments showed that this matrix applies to the nature of physical reality.

In the grand unified theories (GUTs) advanced in the second half of the twentieth century, the quantum vacuum morphed into a physical energy sea that carries the zero-point field, the widely discussed ZPF. In the likewise much-discussed "theories of everything" (TOEs) all of nature's fields and forces are integrated as manifestations of the unified vacuum.

The interpretation of the quantum vacuum in reference to the zero-point field was reinforced when Paul Davies and William Unruh advanced the theory that differentiates between uniform and accelerated motion in the zero-point field. Uniform motion does not disturb this field, leaving it isotropic (the same in all directions), whereas accelerated motion produces a thermal radiation that breaks open the field's omnidirectional symmetry. Since the 1990s, numerous physics experiments have explored this premise. Harold Puthoff, Bernard Haisch, and collaborators demonstrated that the inertial force, the gravitational force, and even mass are consequences of the interaction of charged particles with the ZPF.

Other physics explorations focused on the Casimir force, a phenomenon that occurs when some wavelengths of the vacuum's energies are excluded. This is the case when two metal plates are placed close together. This reduces the vacuum's energy density within the plates with respect to vacuum energies on the outer side. The disequilibrium creates a pressure—known as the Casimir force—that pushes the plates inward and together.

Physics research also focused on the Lamb shift, a vacuum effect manifested in the frequency shift of the photons that are emitted as electrons around the atomic nucleus leap from one energy state to another. The shift is said to be due to the photons exchanging energy with the ZPF.

Further evidence that space is neither empty nor passive has been furnished by the discovery of pressure waves propagating in interstellar space. Astronomers in NASA's Chandra X-ray Observatory found a wave generated by a supermassive black hole in the Perseus cluster of galaxies 250 million light-years from Earth. The wave has been traveling in the vacuum for 2.5 billion years. Its frequency corresponds to the musical note B-flat, but human ears cannot perceive it: it is fifty-seven octaves below middle C (which means that it is more than a million billion times deeper than the limits of our hearing).

The universe proves to be a plenum of in-formed energy. The presence of this active form of "in-formation" is evident at all scales of magnitude and complexity. It is evident already in the bonding of molecules of water. This bonding is essential for the emergence and sustenance of life. Living organisms are about 70 percent water. The properties of the water in the organism make life possible. The bonds between the hydrogen components of H_2O molecules are more than ten times weaker than ordinary chemical bonds. Because of the stretching of the molecular bonds between hydrogen atoms, every drop of water is constantly forming and re-forming, adapting within a living organism to the requirement to maintain the system in the living state.

Vast energy is conserved in systems of bonded atoms, molecules, and the myriad chemical and biochemical structures created by them. The precise value of this energy is not established. If it is as great as physics' Standard Model indicates, not only would distant galaxies recede (as they do), but all galaxies, and indeed all stars and planets, would fly apart. Space would be nearly empty in the vicinity of Earth. The universe would expand like a rapidly inflating balloon.

The Akashic in-formation field

A universal space-and-time-transcending in-formation field is required to account for the kind of connections that create coherence in systems. This coherence, as we said, appears at all scales of size and complexity, and in all domains of nature, from the micro-domain of quanta, through the meso-domain of life, to the macro-domain of the cosmos. Just as electric and magnetic effects are conveyed by the EM field, attraction among massive objects by the G field, and attraction and repulsion among particles in atoms by quantum fields, so universal connections are conveyed by a universal in-formation field: the Akashic Field.

The Akashic Field is a fundamental nonenergetic medium underlying things and events throughout space and time. This tenet corresponds to a time-honored insight of the wisdom traditions. The Akasha, as other spiritual phenomena, is real, although not being perceivable by eyes and ears.

In the past five thousand years, countless spiritual masters claimed to have perceived a field or medium that connects things. This is part of the spiritual vision proclaimed by seers from the ancient rishis to modern sages such as Sri Aurobindo. The Hindu seers claimed to perceive the Akasha through spiritual practice—the disciplined spiritual way of life we know as yoga.

The maverick genius Nikola Tesla adopted this vision. He spoke of an "original medium" that fills space and compared it to Akasha, the light-carrying ether. In his unpublished 1907 paper "Man's Greatest Achievement," he wrote that this original medium becomes matter when prana, cosmic energy, acts on it. When the action ceases, matter as such vanishes, and the universe returns to the Akasha.

Today, over a century later, the Akashic concept is revived. Its plausibility is reinforced by theories promulgated at science's cutting edge. More and more scientists realize that space is not empty, and what has been misleadingly called a vacuum is in fact the field that interconnects events throughout space and time.

There is much to be discovered yet about the properties of the subquantum plenum, alias quantum vacuum. But we already know that it is an information field that transports photon waves (light), as well as the density-pressure waves that replenish the energy lost by atoms and solar systems and decides whether the universe re-contracts to quantum dimensions or expands infinitely in cosmic space. The universe has an Akashic dimension: the dimension the ancients called the Akasha.

The Akashic biofield

Space-and-time-transcending nonlocal connections have been found in all domains of investigation, including observation-based physics, and not only spiritual phenomena. They have come to light in living organisms as well. Superfast, multidimensional correlations within the organism turned out to be a precondition of maintaining complex biological organisms in the physically improbable state we call living. The staggering amount of chemical and physical reactions that take place in the living organism could not be fully coordinated by limited and relatively slow biochemical signal transmissions. One of the most basic functions of cells—communication with other cells in the organism—calls for more information and faster information-transmission than any hitherto known channel could provide.

Through quantum correlations, cells create a coherent field throughout the body. This "Akashic biofield" supplements the biochemical flow of information in the organism with a multidimensional quasi-instant information flow. This is needed to ensure the coordinated functioning of the organism as a whole.

Quantum correlations are not confined to the physical bounds of individual organisms; they extend into the environment. Through its Akashic biofield, the organism is in constant multidimensional communication with its life-supporting environment. This communication transcends the organism's immediate milieu; it involves quantum effects and creates communication with systems and organisms whether they are proximal or distant. In the final count, we can affirm that quantum correlations connect living organisms throughout the web of life on the planet.

In-formed evolution in the universe

The in-formation that links and governs the relations of particles, and of the complex systems constituted of ensembles of particles, creates a definite direction: this is an evolutionary vector. Evolutionary processes are oriented toward creating ensembles that constitute nonrandom systems that form coherent wholes.

The nonlinear and intermittently chaotic, but persistently unfolding evolution of physical, biological, and astronomical systems is evidence that the universe is not a domain of material particles moving in passive space and time. It is a domain of complex evolving systems, oriented by universal "in-formation." This confirms an ancient intuition of who we are and what the universe is. We are *NOT* biochemical machines inserted in an impersonal and passive universe, inexorably running down to cosmic heat-death. We *ARE* evolved clusters of finely tuned vibration, integral parts of an ocean of vibration in a universe that appears to us as an information and memory field evolving to greater and greater complexity and order and manifesting higher and higher forms and levels of coherence.

PART 2

LIVING AND HEALING IN THE AKASHIC FIELD

LIVING IN THE AKASHIC FIELD

Kingsley L. Dennis

*The future enters into us . . . in order to transform
itself in us long before it happens.*

—RAINER MARIA RILKE

An introductory note

Today a monumental transformation is underway. The future is transforming itself in us, as the above quote from Rilke suggests. New vistas are opening up, among them the vista that allows us to perceive fundamentally new insights upon the nature of reality, the cosmos, and our place in it. It is no exaggeration to say that humanity is at the beginning of an unprecedented era.

This is a true insight, but insights alone are not sufficient. What is required now is that we take this new vista and *arrange, assimilate,* and then *apply* it. By this, I mean that the new perspective needs to be arranged into coherency and comprehension. Into a vision that can be shared and comprehended by others. It has to be brought down from the realm of insight and investigation into the arena of the pragmatic. In this way, the new understandings we have of the cosmos and of our relation to it can gain meaning.

We then need to assimilate and process this new understanding to make it real for us—to truly grasp its implications. That is, to internalize what the emerging perspectives can bring and not merely to leave them as intellectual segments. The emerging knowledge needs to become whole and integrated—needs to imbue us with its significance. This is what I mean by assimilation. We must absorb and digest what the implications are for shifting our perspectives to align with the new view of life in a fundamentally interconnected and intelligent cosmos. Finally, to benefit from this knowledge and understanding, we shall need to apply it practically to our lives. If we cannot make tangible and workable this new vision we have gained, how can it truly benefit our lives—our future development as a species upon this planet?

In these changing times—times of uncertainty, disruption, and, above all, transformation—we are in need more than ever of narratives that can help us navigate alongside an emerging world full of new understandings, insights, and opportunities. To put it simply, we stand collectively now at the advent of a totally new paradigm that will literally *shift* our world into a new era. It is this monumental shift that concerns this book, and this chapter.

At the outset of this book, Ervin Laszlo told us that the search for a better life and a better world is a long-standing quest. True, and now the next step is to live it. If humanity is not able to integrate into life the new understanding that arises within the unfolding journey, then what is the point? As an intelligent, sentient species, humankind is meant to adapt the new perspectives it gains into developing itself along both an external as well as an internal path.

As our understanding of the world unfolds, so do *we* develop as human beings. There is no isolationism; each is bounded by the other. And like a feedback loop, the more understanding we receive, the more it pushes forth our perceptions onto ever-greater understanding—like water rushing through a vortex and widening it. It is when we pull back, out of hesitancy or fear, that we stem the onward flow of our growth and realizations. We have now come to understand that reality as we know it (materially and energetically) exists as a totally interdependent matrix—a unified whole.

As Laszlo explained, all life exists as part of not only a participatory cosmos, but also of a sensitive cosmos—of a cosmos sensitive to our presence, our participation, and our projections. And the more that science delves into the structure of the cosmos, the more those structures reveal not only external forms of connectivity and correspondence but, more remarkably and importantly, an unseen yet underlying unified coherence. Just because a thing cannot be seen does not mean it does not exist. Modern humanity has turned away from the romantic era of metaphysics, as Laszlo put it. And yet the metaphysical background of the world persists and even pushes through into the physical everyday world that most people inhabit. First through insight, vision, focused imagination, and now scientific exploration, it has been revealed that a much more "real" world exists as a foundation, or source, for the surface—or "explicate," unfolded (using Bohm's terms)—physical world. What is now known about reality is that it is a mix, an integral blend, of the tangible and the intangible.

The physical reality of the world is now better described as unfolding from an unseen (implicate) metaphysical background. This manifests a physical reality that is navigable and solidly coherent, yet at the same time "glued" or held together by a unified field of energy that is vastly more coherent than the physical. As Laszlo stated in the first chapter, there is more to the universe than its physical manifestations. It is, he noted, a domain of instant universal interaction, of complete and infinite memory, and integral evolution toward oneness and coherence. It is therefore only natural that in response to these new discoveries, our technologies should also, by default, be applying those correspondences to manage relations between the tangible and the intangible to make our world a better place. These are the issues that this chapter addresses one by one.

The Akashic program

To use modern computer terminology, everything within the Akashic Field is part of the same program. Each aspect, or manifestation, is "hardwired" into the collective source—the mainframe, let us say. Every area of the program is intrinsically connected to every other part. Each component, or consciousness, also ultimately depends upon the other to retain the wholeness of the program. Likewise, atoms and particles collaborate and share. This is how they form what appears to us as "matter." They interrelate, intimately, with an underlying drive, which we may call purpose. If they did not do so, no "solid," seemingly material world would be possible. Collaboration and interconnectivity are thus the very basis of existence—energetically as well as materially. This can be called a form of mutual symmetry—a dance or exchange of energy and information that functions through balance, coherence, and resonance. These are values. And it is to shifting values and their meanings that I now turn.

Akashic values and the new technologies

Contemporary science has now recognized that the world humans live within exists as part of a nonlocally interconnected reality. As such, features of this underlying reality are now seeping into the denser, physical world. One example of this is in the shift now occurring in how social infrastructures and modes of human connectivity are moving from hierarchical structures to decentralized and distributed networks. These new modes, accelerated to a large degree through our communication technologies, reflect the current understanding of the Akasha paradigm. These material shifts are corresponding to how we are now assimilating subtler levels of human interrelatedness, and how we are speaking and articulating these new arrangements. In recent years, people have increasingly spoken of connection, communication, decentralization, networking, and collaboration. These are the signifiers of an integrated, holistic view of life.

In this context I have proposed that globally we

are gradually witnessing a transition from one set of "C-Values"—Competition ~ Conflict ~ Conquest ~ Control ~ Censorship—toward a new set of C-Values: Connection ~ Communication ~ Collaboration ~ Consciousness ~ Compassion.[11] What this implies is that there will be an increasing correspondence between how energetic and material flows shall operate.

Prior to this book, Laszlo laid out his understanding of how reality operates—its underlying fundamental patterns, drivers, and the correspondences between energy and the physical reality that appears as matter. These latest findings literally turn our conceptions upside down. At the core of our existence is an energetic pattern of unity. As this grasp of essential unification seeps through our perceptions and perspectives, this will affect how people behave. That is, how we *do* things and the *way* we do them. Our outer circumstances will begin to reflect not only our inner realities but also the reality of the cosmic order. This may sound like a tall order, as they say; yet, the greater is reflected in the smaller, and the smaller contains the seed of the greater. Or, to put it another way, "As above, so below," as the Hermetic maxim goes.

These convergences are now coming to the fore. The connectivity that has always innately bound us within the cosmic reality matrix is now emerging to form tangible relationships as part of our external reality. These new arrangements will create a different style of human ecosystem, and this is likely to be shaped by how we make use of the new technologies. Technology is not something *beyond* us or separate from us. As I express here, technology has emerged as an extension of the human being—a projection of our intention to interact, interface, and influence the environment around us *as if it were a part of ourselves.*

Technologies, I posit, do not arise among human civilization by chance; rather, they reflect the state of human consciousness and our ability to grasp and perceive our sense of reality. Some technologies may even arise in accordance with our state of readiness to understand them—they are a representation of ourselves and reveal our perspectives and insights into how the reality matrix operates. What this suggests is that technologies that are designed to enhance our relationship to the world have an often-overlooked connection with the state of human consciousness. Sometimes this innate relationship is out of balance, is lopsided and incongruous, and results in technologies of destruction (such as in warfare). Other times, the relationship is more aligned and results in creative innovation that aims toward the betterment of human life upon the planet.

The previous mindset or consciousness—sometimes called the industrial mindset—was one that viewed the physical, materialistic aspect of life as dominant. This was a consciousness of acquisition, possession, ownership, and ultimately control. It was all about who had the hardware, and the power to control the hardware over others. It was an age that flourished on patents and copyright, and restriction and centralization. It was all very tangible, and solid, and could be seen, felt, and known. It was about communications via visible aboveground cables, and everything was attached and contained in the grid—in the physical matrix. Then technologies started to change: The cables began first to disappear beneath the ground (or sea), and then to disappear altogether as wireless and satellite became main commercial and civilian channels. The leads between the keyboard, mouse, and monitor vanished, too. Things started to connect in nonvisible ways; and they got smaller, too. Then instead of just the computer we had multiple devices to connect to the ethereal ("Where exactly is it?") Web. And then our technologies became increasingly distributed and decentralized as networking became the dominant paradigm and way of operations. All that was solid was now melting into air (to paraphrase Marx). Also, the spectacular rise in global communication technologies—internet-enabled devices, digital platforms, social networks, et cetera—reflected a new form of participatory consciousness, especially among younger people. With this change emerged also a shift in human consciousness as if, to quote Rainer Maria Rilke, the future had already entered into us.

Visionary thinker and architect Buckminster Fuller coined the concept of "ephemeralization" to express how technological trends were shifting ever closer toward the ethereal/ephemeral, as hardware dissolved into software and information (energy). We are now witnessing how unseen information (such as code) is shared and swapped around the world like invisible electricity—the new Promethean fire. Human civilization, in line with the Akasha paradigm, is transforming from utilizing heavier materiality toward lighter, more subtle forms of connectivity and functionality. It is not happening uniformly yet, although the trend is strongly noticeable in how once less technically developed nations/regions have been leapfrogging over heavier phases of technological development. Life on planet Earth is about to increasingly come together through the subtle energies and channels of communion. This new phase builds upon the digital revolution and is characterized by a more ubiquitous and mobile internet, smaller and more powerful sensors that are cheaper, and artificial intelligence and machine learning. It marks the fusion and interaction of technologies across the physical, digital, and biological domains. This corresponds with the findings of contemporary science that declare the physics of the universe is information and energy. The world as it manifests is based on arrangements and configurations of, as Laszlo calls it, "informed energy." These configurations are now being utilized in a very real physical sense within the external world of human civilization. This is the revolution of the emergence of a completely unique ecosystem upon the planet, a system that aligns with the understanding of the Akasha paradigm. The *Akashic life* is a fusion of external and internal correspondences.

AKASHIC LIFE: A NEW PLANETARY ECOSYSTEM

There is another world, but it is in this one.

—PAUL ÉLUARD

The first Industrial Revolution lasted roughly eighty years; the second Industrial Revolution is generally dated as lasting around fifty; and the third Industrial Revolution has lasted just thirty years. Within this rapidly unfolding scenario a profound uncertainty exists in how emerging technologies will be adopted and developed, and how this new paradigm will recalibrate the social life of human civilization. One thing is becoming clear, however, which is that our technological devices will become a part of our personal ecosystem, responding to us and our needs like an extension of ourselves. These "personal extensions" will leverage the pervasive power of digitization and information technology in ways unprecedented and in many ways as yet unperceived. Within this confluence of emerging technologies will be artificial intelligence (AI), robotics, the internet of things (IoT), autonomous vehicles, 3D printing, nanotechnology, biotechnology, materials science, energy storage, and quantum computing. The intangible internet within us is beginning to merge with the tangible products of our everyday world.

The embedded environment is increasingly communicating with us by the streaming of information. Our personal modes of access are shifting to reflect the *receiving* and *sharing* of information, a form of communication from us to an increasingly responsive environment. This greater participation with information flows is representative of the *Akashic era* we are moving into, as the tangible melts into the intangible. Our digital identities, traces, and networks will become as much a part of our lives as our physical footprints. This is in fact a more natural reflection of how energy operates in the cosmos—it is already all around us, yet each localized part *streams* the transmission from the nonlocal field (as explained by Laszlo).

The internet of things (IoT), as it is now often called, is

envisioned as creating a relationship between products, services, places, and people to form a cross-connected platform. It is a future where nearly every physical product could be connected to a pervasive, ubiquitous communication infrastructure, and trillions of sensors will allow people to perceive and navigate their environment in new ways. In other words, an ecosystem will be established that merges the physical, digital, and biological. There is as yet no name for this new ecosystem. It has previously been referred to as the emerging Global Brain; yet this fails to comprehend the overall physical immersion. It is more likely to represent a central nervous system for the planet.

Our technologies are fast merging—or rather *streaming*—into our environments, becoming increasingly more seamless with the human body-mind complex. This shift in the evolution of our technologies has come about almost naturally, and we have seldom questioned why it is going in this direction. Yet if we consider how consciousness and the information of the Akashic Field operate, then there is a pattern. Our reality matrix *receives* the creative energy-information from the unified Akashic Field, just as consciousness *streams* into our minds (as a localized receiver within the nonlocal field). Thus, for our technologies to shift into this mode represents an overall, integral pattern. In this way, technologies are very much a part of the Akasha paradigm and are neither divorced from nor antagonistic to it.

These emerging patterns are facilitating a shift from a culture and mindset of acquisition toward one of participation and sharing. As previously stated, as the Akashic life unfolds, we shall see a greater influence of the values of connection, communication, collaboration, consciousness, and compassion. This is an energy that supports not control or ownership but collaborative participation. Robert David Steele has referred to this mode as the Open-Source Everything.[12] A dynamic and creative energy is forming a new ecosystem as it moves from a linear-energy system to an energy that is integrated, restorative, and regenerative—and which moves not in straight lines but in waves, networks, and circles. These energies are pushing to form new arrangements and "Akashic relations."

Akashic relations

As Laszlo emphasizes here as well as in his previous books, the universe as a whole manifests fine-tuned spatial and temporal connections. From these connections is established a level of correlation and correspondence between the bodies that manifest physically. Likewise, as the "Akashic ecosystem" emerges and settles into arrangement, new symmetries of correspondences, information sharing, and social relations have arisen and will continue to arise. Social and cultural urban geographies will be recalibrated and repositioned. After all, ecosystems are innovative and creative and need to be continually nourished and reconsidered. Within this ecosystem, the digital-physical connection will be at the forefront. It is this mix of material-informational relations that shall constitute a reproduction of the "Akashic life."

Similar to how our manifest reality is coded from the unified Akashic Field, so, too, will code become an integral part of the new planetary ecosystem as we will increasingly exist alongside ethereal-like algorithms that code and recode the environment as we move through it. Physical-digital spaces will become more adaptive and customized to our needs and be not only urban spaces. Nature will become a more integrated component in the emerging ecosystem as more greenery (plants and trees) are adopted into buildings, our living-work environments, and monitored to ensure optimum growth and health. The integrative, holistic "Akashic view" is that Nature is not something separate from us, but rather that we are in communion with it; as such, Nature is a collaborative participant within the grander, inclusive ecosystem of relations.

Things external to us may not appear to be in connected relations, such as has been the case with technology, as these take time to arrange and settle themselves into a more harmonious relationship. What is being experienced now in various situations around the world are shifting patterns of energies, as top-down hierarchical flows are moving into more decentralized, circular flows. In this mode, life will appear to be moving at an increased pace, as these decentralized flows facilitate a more fluid interaction and unfolding of events. The importance of connection is to shift toward the greater importance of being relational. That is, point A and point B may be connected, as in the past, yet this may still be a linear type of relation and energy. Rather, both points A and B will participate in a relational way; that is, across multiple levels and in nonlinear forms. Humanity in the Akashic life will shift into more relational connections. It could be said that a "feminine" type of energy is entering more into global systems.

The feminine impulse (which has no connection to gender) has already made inroads into our global institutions and technologies. It has been the energy behind the restructuring and recalibrations now unsettling our societies. This has been borne out in the shifts now taking place as many top-down hierarchical systems are transitioning to bottom-up, decentered, and distributed systems. The most prominent example of this is in our global technologies of communication. As an example, our modes of communication have shifted from one-to-one (such as television) to many-to-many (digital communications, such as the internet). Our vastly expanding digital world is more than a communication device, more than an "inter-net"; rather, it is a mirroring of our own cosmic *inner-net*.

Humanity has always been a part of symbiotic life on this planet. Nothing—no species—exists in isolation. Symbiotic humanity is now extending this relationship into a technological partnership. It is a partnership of *technoosis*,[13] as the *tech*nosphere and the *noos*phere create another interconnection—or inter-relationality. It can also be expressed as matter and mind merging and coalescing into a unified, coherent Akashic field of communication. This transition calls for a convergence of biological, technological, and ecological concepts; this also reveals the need for more prominent feminine values and relations.

These feminine values and relations are essential in how our complex and decentralized networks are recalibrating the ways to connect and communicate. New initiatives, innovations, projects, friendships, and relations are emerging from this interconnected technoosis. The new multiplicities are undermining the once-dominant top-down hierarchical systems (which we may denote as the masculine energy). The new collaborative spaces are all about multitasking—from share economies to open-access information. Our global platforms are increasingly becoming a place/space for such issues as human rights, education, health care, child care, welfare, the environment, et cetera. These mounting issues, as well as the manner of how they are multitasked and openly discussed, all belong to the new Akashic values of collaboration, connection, communication, consciousness, and compassion. Physical institutions and systems are responding to these shifting modes and, as such, are displaying signs of uncertainty and breakup amid the rearrangements.

The new planetary ecosystem is forming as a biological-digital-technological matrix akin to a new mode of a living organism that is becoming entwined through the exchange of ideas, information, innovation, goods, and services. Some of the trends we are likely to see emerge within this Akashic ecosystem are wearable interfaces/internet; connected homes; smart cities; driverless cars; robotics/robotic services/drones; quantum computing; digital currencies (blockchain); 3D printing/manufacturing; sharing economy; digital identities; personal devices (with supercomputing power); and access to the ecosystem as a fundamental human right.[14]

All of these changes mark a shift to new models of exchange, away from ownership and toward access. That is, moving away from clunky hardware toward more subtle forms of software (the flows/relations rather than the physical infrastructure). The ecosystem will be constantly nourished from user-generated content, representing increased participation from people within those flows and connections. These forms of Akashic relations will naturally facilitate individuation (rather than self-centered individualism) and create new forms of community and belonging.

Access and use of the physical-digital ecosystem can help in dissolving and breaking down social divisions as individuals realize that they are empowered actors as part of an increasingly decentralized and collaborative system. The transcending of time and space limitations will further encourage and support mobility. However, increased connectivity does not necessarily create immediate tolerance; there will be displacements along the way as the new geographies attempt to be defined. Yet it is almost certain that increased physical and digital mobility will have a greater role and impact upon societies and on human civilization, restructuring cultural norms and social behavior.

There will also be many continued inequalities in the world as we undergo the transition to an age that reflects the Akashic values. Yet these inequalities and human brutalities are not the consequence or fault of our technologies—they are *human faults*. They stem from human thinking and the older values of greed, power, and control that represent the old paradigm system. No great technological marvel will alter these human traits—only a shift in human consciousness. And this is already occurring within each new generation coming into the world. The "Akasha shift" may not sit kindly, or easily, with many of the older minds of the older generations. Many "older-minded" people are feeling overwhelmed by this sudden rush of change and recalibration. The newer generations are being born into a new planetary ecosystem that will feel natural to them, where old boundaries and frontiers have been rearranged.

This new pervasive ecosystem is not without its pertinent risks and dangers, as it is still a recent and innovative playing field. As such, many new players—from government bodies to rogue and extremist organizations—will continue to utilize and experiment with the new digital platforms. There are significant opportunities as well as risks as the culture of networking and the sharing economy does not come about without its opposing shadow of oppression and suppression. A different world is emerging—being birthed—and it needs to find its place. Values are being shifted, new modes and ways of doing things are causing disruption, and many things—as well as many people—are on the move. Many aspects of our societies are shaking off the older energy relations and recalibrating for a new "Akashic life" to unfold. It will not be a totally smooth process. There is a clash of ideologies-mythologies occurring right now during this transition period. Yet, greater instability is also a sign of incredible change on the horizon.

Akashic identities

Rapid and innovative change is an essential and necessary component of the emerging life in the Akashic world. The new ecosystem coming into arrangement is not only impacting *how* we do things but also *who* we are in this changing reality. The notion of human identity is experiencing further shifts in how we relate to our fellow humans, and how we understand ourselves as part of the fluid environment. As our reality becomes increasingly augmented, it will force us to question our species identity as well as our role and future upon the planet. Debates concerning the quintessential features of human identity are set to become central to the Akashic life—"being human" will become a delicate and incendiary subject for some people.

Those new generations who will be born into a fully digitized world will not have the same issues as those of us forced to adapt to it from earlier perspectives. Many will resist the changes; this always has been and always will be the case. Revolutions occur across many different spheres, yet when they come together in a grand confluence—physical, digital, mindful, and spiritual—the timer has been set for some radical unfolding and evolving.

The incoming phase of change that the Akashic life brings to human civilization will compel us to reconsider, and perhaps redefine, what it means to be human. The seeds for this were planted decades before, especially in the late twentieth century. It will compel us to deeply consider and understand, perhaps through much soul-searching, what it means to be human. We are only now, as technology becomes a predominant part of our lives, beginning to consider how these emerging connections and relations are also affecting our inner selves.

The changes associated with the emerging Akashic-life ecosystem will stimulate the strengthening of inspired intelligence; that is, one's instinct and conscience in the search for meaning, purpose, and direction. Inspired intelligence expresses itself through consideration, understanding, and imaginative lateral thinking outside compartmentalization. This is the fluidity entering and affecting many incumbent social and cultural institutions and is likely to increase in the years ahead.

Psychologists are already recognizing the rise of empathy among many of the younger generation as these youngsters are able to relate to people, often strangers, across the globe who are in difficulties or experiencing challenging times. This sense of human solidarity, a feeling of compassion for others, is growing stronger within the minds and hearts of the young generations as they feel naturally connected to those from beyond their localized communities. Identities are becoming more fluid as people recognize themselves as being planetary citizens rather than relating solely with national identities. These shifts indicate new patterns in terms of how, as a human community, people live, connect, communicate, collaborate, understand, produce, and create value and meaning in their lives. And in the Akashic universe, life spans are set to increase.

Life spans within the unfolding Akashic reality are set to increase around the world, affecting younger-generation learners and their skill base. Living longer, healthier lives suggests multiple career paths that utilize varied skill sets rather than one career path. This compels younger learners to be open and adaptable to new skills and opportunities. Another factor to take into account is that as people increasingly live and work alongside a technologically enhanced environment, they are more likely to focus on the skills that are uniquely human. Rather than dehumanizing us, such smart environments can refocus our thinking upon the skills, assets, and inherent abilities that make us human.

The new Akashic ecosystem is set to transform the way people communicate, produce, and share their thinking and creativity. This will especially relate to a media ecosystem that consists of social media, video production, gaming platforms, augmented reality, and citizen journalism. This landscape, which is ushering in new modalities of communicative language, will find itself serving as a critical response to social, political, and cultural behavior. Young people especially are strong media-content generators, and they have the potential to offer innovative projects as well as producing multimedia commentary upon social events. They represent a powerful new wave of citizen journalism that has the power to penetrate beyond the hypocrisy often provided by mainstream news outlets. A strong online presence in this new media landscape places new demands on cognitive and attention skills relative to these collaborative platforms. These social media platforms are likewise strongly indicative of how structures and organizations are moving away from the older hierarchical top-down forms and into decentralized and distributed relations. This is affecting how people are interacting and communicating both at local as well as global levels. These changes across our diverse cultures and societies are already influencing and shaping the minds of the younger generations, who are displaying a new range of skill sets, such as social intelligence, sense making, adaptive thinking, cross-cultural awareness, and new media literacy. We will now take a brief look at some of these "Akashic skills."

Akashic skills

Social intelligence implies an ability to connect deeply with others, both in physical and digital relations, and especially in extended networks. It also suggests a greater awareness and sensitivity to interactions across national, ethnic, and cultural groups. Due to the nature of living in an interconnected world, new sets of social skills are required that demand a sensitive understanding to diversity and difference. Making sense not only of our environment but also of the meaning behind words, expressions, and social behavior is one of the features that make us human.

As a complement to sense making, adaptive thinking is also a feature of the current transition within human consciousness. Old rules and ways of thinking no longer apply as the world undergoes a radical shift in its systems and old-paradigm models. This broad-based integral thinking is an essential part in developing cross-cultural awareness. This is sometimes referred to in psychological terms as situational awareness or situational adaptability. This is an important skill, especially for the younger generations, as it is about embracing holistic thinking and behavior that transcends linear patterns.

New forms of media literacy also allow people to participate across a range of multiple platforms. This includes the creation of their own video/media as communication continues to become highly visual in our digital cultures. We are seeing the growth of more visually oriented practices, especially as tools of augmented reality become more common. In this way, the minds of the younger generations growing up within the Akashic paradigm are likely, through cultural stimuli, to become more highly developed in the visual, and hence right-brain, perspective. Again, greater activation of right-brain awareness allows for creative expression, and for seeing the "bigger picture" outside the classical categorical thinking.

In short, we have entered the beginnings of a new era when great changes are occurring throughout varied sociocultural systems as human life adapts to an integrally connected world that is itself a reflection of the underlying Akashic Field, as expressed by the latest scientific findings. What this also indicates, and which is central to this, is that we are now being presented with a new perspective on human consciousness.

Akashic consciousness

Human consciousness is going through an unprecedented transformation as it recalibrates to process the new understandings and perspectives. This new information provides what we could rather crudely refer to as an "upgrade." It would be fair to say that what is being experienced now is on a par with what occurred as human consciousness shifted from a "flat earth" to a "round earth" perspective. That turnabout changed everything. It is no understatement to say that human life on this planet is on the cusp of a revolutionary shift in perceptions that will mark everything that occurs afterward. Alongside this shift, I propose, will be a strengthening in the recognition and expression of one's inner authority. And this will manifest alongside a rise in what is termed *instinctual intelligence.*

Rather than acting from acquired information (based on educational conditioning), individuals with "Akashic consciousness" will increasingly trust an internal sense of what needs to be done. This form of gnosis is more in keeping with the harmonious relations of systems that act upon coherence and balance. Also, as people learn to trust their inner instinct, they are increasingly seeing through the fog of cultural conditioning. This is important to stress as much of the old-paradigm thinking has been propagated through forms of conditioning. Thus, people are going to become increasingly aware of social and cultural propaganda. As such, people will be less susceptible to the manipulations inherent in the institutions of economics, politics, and health, as well as falsehoods and improprieties connected with religious doctrine and pseudospiritual teachings. The veil will begin to lift as forms of transparency will divulge the "wizard of Oz" pulling the levers behind the curtain.

Many incumbent belief systems are likely to fall away as the Akasha life unfolds. One of these is the belief that we are no more than a physical human being and that there are no other realms/dimensions beyond the earthly experience. This will then lead to revised interpretations of ancient wisdom as people awaken to the potential for change that is within each of us. What this will indicate is a recalibration of understanding and a realignment of spiritual systems for a more enlightened consciousness. The Akashic consciousness is one of tolerance and respects the spiritual core of a person, and personal development and self-evolution are essential values. The Akashic life brings forward the role of wisdom within human institutions.

All these changes and shifts in our lifestyles, institutions, and social systems will not happen overnight. Like ink dots spreading across blotting paper, they will occur in phases until eventually the paper has crossed the tipping point and flipped into a new color. Already it is seen how many (although not all) young people who are to take up the mantle for advancing change upon this planet are expressing themselves in a way that embraces what we refer to now as the Akasha paradigm. Theirs promises to be an exciting world full of tremendous possibility, potential, and opportunity. It is also a time for great courage, self-belief, and selflessness. We are witness to a world in transition.

Whether we choose to manifest things in life, facilitate for others, or nurture others—each involves conscious participation, collaboration, and cooperation. As Ervin Laszlo has shown, cooperation, which leads to coherence, has been the constant and core driving impulse behind evolutionary trends throughout the universe.

AKASHIC RESONANCE AND PLANETARY COHERENCE

If we allow ourselves a momentary grand sweep of history (as in a different context this book has already undertaken in the introduction), we will see that the checkerboard shows the rise and fall of countless civilizations, empires, and cultural manifestations that have reflected also the shifts in human perceptions and consciousness. How we *see* the world, and our place in it, has always influenced how, and to what degree, we participate in and create the world around us. And until recently the consensus has been to view the world as external to us: separate and fragmented. In the pre-Akasha paradigm, values of conquest and control have dominated. Powerful empires have sought to create, as far as was possible, their idea of a unipolar world. Yet no empire ever truly succeeded in this endeavor. The fundamental energy of sustainable growth requires any groupings—such as human systems—to seek greater stability and coherence.

The drive toward achieving greater levels of coherence within and between human societies would lead toward ever-increasing global interconnectedness and interdependence. And the ultimate scale-up along this trend would be a planetary civilization. Within this perspective, it can be seen how the Akashic life represents a developmental impulse toward a planetary civilization. The age of one empire alone dominating the world is at an end. Our present multipolar world reflects a level of deep interconnectivity between the dominant and also not-so-dominant nations, states, and regional blocks.

In this study of the Akashic era we see coherence coming from the bottom up, through the emerging Akashic ecosystem. Humanity is a vehicle, a channel, for *extending mind* across the earth; that is, for embracing the planet like a membrane, a living skin. Just as the skin is the largest organ of the human body, the living skin of the planet will be the integrated physical-digital ecosystem that is driven toward coherence through consciousness. Our developing perception of the interconnection between parts of a whole serves as both an expression of coherence as well as a driver toward further coherence. As stated, how we *see* the world also influences how we participate in and create the world around us.

The current drive toward social coherence across the planet emerges first through the individual consciousnesses of people. And it spreads through local and global systems as part of the emerging and growing networks of interconnection. This drive, toward coherent order, underlies our reality, as shown by the Akashic paradigm. The impulse will push toward greater coherence uniting physical (tangible) and digital (intangible) systems, with human consciousness as a key driver. Just as our technological computer systems use the binary code of 0s and 1s to form a communicative whole, so does this reflect the emergence of the tangible (1) with the intangible (0). Again, this is a formula of the Akashic life that represents the integral wholeness and merging of so-called matter with informed energy.

The world we experience reflects a grander underlying reality based on the Akashic laws. Also, as human beings we each interact with the world differently, because we *perceive* the world differently. In interacting differently, we each contribute to creating a participatory world. The reality of the Akasha paradigm understands that we exist as part of a participatory cosmos. This knowledge infuses the human condition. To be a human being is to be inherently imbued with a life force that animates us in ways we are largely unaware of. Yet through this animated force we perceive the world around us—it cultivates our worldview, our values, and is the source of our quest for meaning. And a civilization's worldview is its most precious possession. Everything proceeds from this primary perception; it is akin to a collective gaze of wonder—or its alternative, of limitation.

Our fundamental understandings are developed within specific cultural environments. These cultural contexts construct the lens through which we view our life and our sense of reality. Older traditions arose as a response to a different world, within the context of a different worldview, and in order to articulate different dimensions of the human condition. We articulate the essence of the human condition according to our times. Now we are saying that the time has come to articulate the human condition according to our understanding of the Akashic laws. As a species, we evolve in the context of our times and in response to a shifting and unfolding understanding of the universe and of our reality.

Within the Akasha paradigm, the world is becoming an exciting, magical, and mysterious domain once again. As noted, technology is moving from its position as a brute, mechanistic hardware to an almost-seamless, magical part of our fluid reality. The world is reviving its sense of being a *misterium tremendum,* a sacred place to dwell in.

The global body, with its expanding modes of communication, is meshing us into a planetary embrace that is creating, cell by cell, a new species body. As this unfolds, we need to meet this transformation by changing the ways we think; by altering the ways we *do* things; by allowing consciousness and the Akashic connection to flow through us. That is, to manifest the qualities, attitudes, and our presence in the world that will most effectively receive, hold, and transmit this consciousness. The Akashic life of today is a nurturing energy that comes alive *through people*— the appreciative touch, the supportive word, the reassuring glance, that we each can weave into our lives. This is a part of the *living soul* that holds within it the species body. As Meister Eckhart said, "The soul is not in the body; the body is in the soul."

The Akashic life is already affecting us, influencing our thinking patterns and consciousness whether or not we are aware of it. Our perspectives on the world and the universe are dramatically changing since the global disruption of 2020. Most of those who have considered the question of life in the universe have realized that we do not exist as part of a dead universe. Even our sciences, our telescopes, are pointing their attention toward intelligent life in the universe. We are closing in, slowly metamorphosing out of our cocoon of cosmic quarantine.

A part of our transition to an Akashic life is the recognition that the human being is a significant part of an enchanted cosmos. Enchantment has been humanity's natural state for eons. The innate state of humanity is to feel integral to all life. This provided for the integrity of the human psyche. This continuity has only been disrupted for a number of centuries, whereas our state of enchantment has been with us for millennia. It is time we return to that enchantment, and to a reconnection with a source of meaning. Those streams of significance, those waters of wisdom, have always been with us. It only depended upon whether we wished to get our feet wet.

As we come together, increasingly through our systemic connections, we each can bring a spark into the burning flame of the integrally connected species body. This is happening through the ever-increasing numbers of people who are seeking a different way of life, seeking meaning and significance in other quarters. As Sri Aurobindo said, such an age must "be preceded by the appearance of an increasing number of individuals who are no longer satisfied with the normal intellectual, vital and physical existence of man, but perceive that a greater evolution is the real goal of humanity and attempt to effect it in themselves, to lead others to it and to make it the recognized goal of the race."[15] The antithesis of the Akashic life is that which seeks power in the sorcery of psychological control and manipulation.

The integral connection between the inner world of the individual and the external physical world is a synthesis that can provide immense meaning for us. Life must have meaning for each of us individually before we can bring authentic meaning into the lives of others. The Akashic life is a driver toward planetary coherence and is not only for the privileged few to the exclusion of all others. Rather, it is open, available, and waiting for all of us, together, to resonate and find harmony and balance as part of our global regeneration and renewal.

Our ancestors were aware that they lived in an integral, Akashic dimension, where the physical world and the energetic realms existed in communion and correspondence. Just as the Akashic energy is a reflection of the human, so the human is a reflection of the Akashic connection. The Akashic-life worldview accepts not only the metaphysical but also the magical and the mysterious—the magnificent wonder in all and in everything. As we evolve, individually and as a species, so, too, will we become intelligible to ourselves and become capable of recognizing the truer nature of the universe and reality. Then will we also come to understand our participation and role in the Akashic agency.

AKASHIC AGENCY: PURPOSE AND MEANING

We are the eyes through which the universe contemplates itself.

—HENRYK SKOLIMOWSKI

The creation of various cosmologies is an inherent human response as a way of experiencing our given reality. The cosmology we hold as an individual, a culture, a species, is a reflection of our state of consciousness and its access to knowledge. For far too long, humanity has considered itself separate from the cosmos. As Jung once remarked, the human feels as if exiled and cosmically isolated. That was also why Jung noted that human beings do not know themselves. Yet the long trail of human history and civilization is, at heart, also the history of human agency.

The philosopher Karl Jaspers referred to the period from 800 to 200 BCE as the Axial Age. It was a time that, according to Jaspers, new yet similar ways of thinking appeared in Persia, India, China, and the Western world. He indicated also that the Axial Age represented an in-between period, where old certainties had lost their validity and new ones were yet to emerge. The new religions that arose in this time—Hinduism, Buddhism, Confucianism, Taoism, and monotheism—influenced new thinking in terms of individuality, identity, and the human condition. These newly emerging religions and/or philosophies helped to catalyze novel forms of thinking and expressions of human consciousness. Yet, over time, we have seen how they were not wholly successful in developing coherence in a social context. Duane Elgin referred to our present time as the Second Axial Age, in that religions of separation are being replaced by a new spirit of communion.[16] In other words, the world is moving into communion and empathic connection with a living universe. The Akashic paradigm reminds us that there is nowhere else to go and no need to go when the universe already exists within us.

This emerging empathic consciousness that Elgin speaks of is the same that has been put forth throughout the chapters of this book. It is an energetic reality that is in-formed and creatively maintained through integral connection and the drive toward coherence. The more the individuated forms of consciousness connect across planetary networks, the higher will be the perception of this interconnectivity. This shall, in turn, catalyze the drive toward ever-greater forms of coherence.

Through its immutable laws, the Akashic Field is a self-referencing feedback loop: the recognition of our participation in nonlocal consciousness amplifies the capacity to receive it, which further expands our recognition of it, and so on. It is a premise, suggested by this book, that sentient human life on this planet serves a purpose as a driver toward establishing a coherent planetary consciousness. That is, to act as an *Akashic agency* for the expression of consciousness into physical manifestation. Immaterial consciousness (informed energy) can be merged within the manifest plane.

Akashic agency acts as expressing localized aspects of consciousness (from the Akashic Field) in order to increase the overall coherence of consciousness on the planet. Further, this can be made tangible by the "local agency"—i.e., each one of us—becoming aware and participating through our everyday acts of right thinking, right behavior, and right being. What this represents is a shift toward an increase in globally conscious, ecologically sensitive, and balanced individuals who are consciously aware of their responsibilities—which includes their intrinsic connection with others.

We can no longer afford to continue as selfish individuals or an inarticulate mass. We must now take up the mantle of having agency as aware individuals who seek to consciously connect, collaborate, and care about the future. Each of us—as localized consciousness—is a reflection of nonlocal consciousness; and in this way we are also a reflection of each other. As such, no individual lives within a shell, separated from everybody else, but is connected to all by both a tangible and intangible communion: this is our true humanity. As we connect and share our collective thoughts, ideas, visions, et cetera, we will be helping to strengthen the signal of a coherent Akashic consciousness. A planetary consciousness, as expressed through a sentient, individualized humanity, may not only be a real possibility; it may well be the fundamental cosmic purpose conveyed by the Laws of the Akashic Field.

Purpose

It has previously been put forth in this book how the latest science suggests that our reality is coded from beyond the manifest dimension of the universe—from what has been known as the Akasha dimension. As such, our known "reality" behaves in a way consistent with what we know as a holographic projection. In other words, the totality of our reality is *in-formed* by a deep consciousness that is beyond it. The known universe thus acts as a whole, nonlocal consciousness field, of which sentient life acts as a localized manifestation.

It has been inferred through various sacred texts and traditions that the universe (manifest reality) came into being as a way for its Source "to know itself"—*I was a hidden treasure and wanted to be known.* This is reminiscent of *know thyself,* the famous maxim from the Delphi oracle. Self-reflection is one of the prized attributes of self-consciousness—yet how can the whole reflect upon itself?

Self-realization is something we credit to each attained individual consciousness; it is a path in which purpose and meaning are core drivers and potentials. Akashic agency recognizes that human beings are naturally driven by a longing, a purpose; this signifies a connection, a communion, with a larger source beyond the individual. It is not a question of *ascending,* as this signifies a leaving, an escape from; rather, it is a question of *transcending,* of reaching for connection that is beyond the physical self. And this connection to a "source" beyond oneself (to the Akasha dimension) increases a person's sense of inner *knowing,* which the person is then more liable to act upon. If more people in the near future were to act upon their inner knowing, we can only imagine what high forms of social cohesion and advancement would result.

To reiterate, human consciousness is a localized expression of the greater nonlocal consciousness that is the Akashic Field. As individuals, we are animated by access to this greater consciousness field, which is manifested through our own socialized minds and cultures. To put it another way, we are all localized reflections of a grander, unified existence. The greater our individual perceptions and conscious realization, the stronger the connection to accessing the Akashic Field. In turn, the Akashic Field is likewise "in-formed" through the emerging awareness of each of its conscious elements or parts. The realization of living the Akashic life is that we each have a role in bringing the unfinished world into completion. Through our daily living, we each cooperate in bringing the world into existence.

If enough of the individuated forms of consciousness were to develop further their intuitive knowing and conscious access to the Akashic Field, then we would likely be enabling also a planetary consciousness field to awaken into awareness. In this case, we are each a conscious agent of cosmic realization and immanence. We each have a responsibility in our existence on this planet to raise our individual, localized expressions of consciousness. In doing so, we both influence and inspire others in our lives to raise theirs, as well as reflecting back our conscious awareness into the Akashic Field. In this way, we act as both "Akashic agents" of the universe, as well as caretakers for our own planetary development.

The Laws of the Akashic Field show that our reality is not a static state but an active, fluid realm that makes demands on us. As such, we need to step up to our rightful place in the scheme of things and to embrace the obligations and responsibilities that come with being Akashic agents. Part of our purpose is to become aware of our creative contribution to reality and to intensify our connection to the grander whole that in-forms us. We have intrinsic purpose in recognizing ourselves as expressions of a living universe. We can restore the natural harmony and communion of existence by reciprocating the two-way communication.

We can achieve this reciprocation through our own even small acts of conscious participation. The social and cultural change currently occurring across the planet may well be part of this process, in-forming an extended mind and global connectivity. Everything ultimately stems from the unified Akashic Field—all science, all human expression, is a form of integrated energy and intelligence; and with each advance we move a step closer to communion with the unified source field.

The Akashic reality animates the expression of consciousness at the individual, collective, and planetary levels. We may witness a grand awakening of a collective expression of consciousness upon the earth—and this may well be the purpose for sentient life, as Akashic agents. The hidden treasure that is at the core of our existence wishes to be known—for *us* to be known—by our individual journeys of self-realization. We are not alone: a great planetary future awaits us. Knowing this brings immense meaning and significance to living life in harmony with the Laws of the Akashic Field.

Meaning

Living as part of the Akashic world brings new meaning to the sense of wholeness and integration. Wholeness now means knowing that we are part of everything. We are in everything and everything is in us. Feelings of belonging, connection, and communion—with others, with our environment, and with the cosmos—can stem from this initial understanding of our Akashic reality. If it is verified that all existence is a part of a unified, intelligent field, then everywhere in the world is sacred. There is no separation, or place, that is not in connection at all times with its source field. Your own backyard—right where you are sitting now—can be treated as sacred and in communion. Any person can tap into this source of energy, of inspiration, from any location. Since we exist in unity, the essential connection is everywhere, and always available.

Living in the Akashic Field is first and foremost a knowing, and then a deep acceptance, that everything exists as part of a unity. To nurture and sustain this understanding provides immense meaning to our lives. We can cultivate this awareness by reflecting on the following points:

- All existence is a unity. This unity manifests both through essence and through form. In our earthly existence, we are in form. Yet at all times each physical entity, and this includes the human being, is in touch with a formless unity. A person may only *feel* disconnected from this greater reality. Remembrance and recognition can activate a conscious connection at all times throughout our daily lives.

- Unity may appear as fragmentation in the physical world. This is how bodily senses perceive effects. Just as when light hits water it appears to be deflected at an angle, the fundamental unity is likewise deflected by the medium of physical matter. Yet this is not so. Do not be deceived by the bodily senses; feel and intuit the inner communion.

- Love has traditionally been associated with intimate associates, family, and close friends. Genuine love is unconditional, and this reflects living in the Akashic Field. This love can be extended to strangers who are on the other side of the planet. We can love differently, yet at its core the love vibration is essentially the same. By engaging in openhearted connection and communication with others, we will realize that we share unity through love.

- Humans have flaws. This is a part of our natural existence and what also makes us so wonderful. All flaws create a unity. Do not look at flaws and see them as cracks. See flaws as part of the stitches that hold together a greater, unified tapestry.

- Enlightenment is not at all about reaching a "higher" place. People struggle because they feel they need to reach another place. Every place is here. It is a splintered mind that creates these separations. The

Akashic life recognizes that a person need not be in any other place but here, where the person is. Feeling the unity is not "beyond" or "higher"—it is where you are sitting right now. You don't need to get to anywhere—you only need to be.

- Living in the Akashic Field means there are no spaces of emptiness. There are no holes or gaps, other than the ones we create for ourselves. Fragmentation is a human creation and does not form part of the greater reality.

- Aligning to the Akashic paradigm life is about being receptive to flows. Flows create harmony and balance. It is important to allow oneself to open up to the flows. Do not struggle against what makes you uncertain. Some of the best opportunities come from unexpected areas. Allow these moments to arrive by accepting that things may flow in ways you do not initially understand. Wait, be patient—do not try to immediately block every unknown event in your life.

- To align with an Akashic life means it is important to cultivate balance. Things that may be good for you should not be focused on exclusively at the exclusion of other things. For example, you may wish to be a vegetarian, yet if this is not balanced by a healthy lifestyle overall, then there can still be imbalance. Do not consider things in isolation, but as part of a grander web of balance and correspondence.

- It's okay to allow some of the excessive baggage to fall away. Possessions are often those things that own us. Possessions are not only material objects; they can also be thoughts, grievances, desires, frustrations, et cetera. Every now and again we need to do a spring-clean upon ourselves and take out the excessive and unneeded baggage. Recognizing the Akashic Field is also about recognizing one's own inherent freedom—physically, mentally, and emotionally.

- Try not to be in total control of everything—no one is in total control. That is only a human myth. Un-expected things will happen in an Akashic life—let them occur. Be flexible. Bend in the wind like a willow tree. To bring things into harmony and balance is often more important than to bring things under control.

- Physical exercise is beneficial, yet no one practice is better than another. For example, yoga is no better than walking if you don't feel comfortable in yoga positions. Don't push yourself, or your body, into places that don't resonate with you. Feel what exercise works for you and take it steady. Don't be extreme—be joyful.

- You don't necessarily need to reach a "higher" state or achieve this or that, or be better, be the first. You only need to remember that you have the connection with the unified whole within you. It is almost like being connected by phone. If you feel you need that connection again, pick up the phone and return the call.

- We often get lost in our ways of naming this, defining that, separating one thing from another. These are human-created traits. It is useful to classify things

in life because the external world needs these defini-
tions. Yet the Akashic life does not need disunity.
Keep the energy unified within.

- Living in the Akashic Field is about finding your
truth. This comes from intuition and "gut feeling."
It also comes from listening to the silent voice that
speaks within you. Practice listening to this silent
voice. Learn to know its feel, its texture, its resonance.
Try following it more often. Learn to trust it.

- When we are about to think, or speak, in terms of
"either/or," try replacing this with "and" and observe
how the process, or situation, changes. Akashic living recognizes that inclusivity is a reflection of the
greater unified whole.

- Akashic living is about manifesting coherence in our
lives. To assist this, we must drop our fears. Fears over
identity, status, of "losing our personality," are especially dominant blockages. Fear causes our receptivity
to shrivel up. Access to the Akashic flow means we
observe the roots of our fears and let them go.

- Identifying with one's country, or community, or
team can be a positive identifier. Yet if this shifts into
competition for superiority, i.e., "mine" is better than
"yours," then we foster separateness and fragmentation. To resonate with the Laws of the Akashic Field
we need to shift from feelings of superiority to feelings of unity.

- Coherence and the drive toward wholeness are embraced by the values we align ourselves with. Try
to remember, in everyday life, that the Akashic

C-Values of connection, communication, collaboration, consciousness, and compassion will always create greater receptivity than the old-paradigm C-Values of competition, conflict, control, conquest, and censorship.

Summary. Humanity has come a long way on its journey, from ancient wisdom traditions to the latest findings of contemporary science. Ervin Laszlo notes that the connecting of contemporary science with ancient wisdom has often been considered a purely intellectual quest to achieve meaning and understanding. While there is indeed a great measure of intellectual satisfaction in this undertaking, as Laszlo notes as well, it is not its sole value. The guidance we can derive from this understanding is a major value as well. The path on which humanity finds itself is not to be "mentalized" (or processed through the brain) so much; but rather, embodied and processed through immersion. In other words, we are seeking a better life in a better world through resonating with the Laws of the Akashic Field. Laszlo is on the mark when he says that today this quest has reached a whole new dimension. It is, as he says, a quest for a better life in a better world. A quest for a future-in-potential the likes of which we are currently unable to perceive or even imagine. Yet we are moving toward it.

What the Akasha paradigm reveals to us is that as we increase our awareness, understanding, and perceptions, we correspondingly develop our connection with the intangible dimension of reality. As this correlation advances, the emergence of informed energy as the foundation of what appears to us as matter simultaneously advances.

Living in the Akashic Field is not only about how we learn to navigate through new and emerging social-cultural-digital spaces, but also, significantly, how we adapt to an expanding awareness in human perceptions. Some of these questions are answered through the contributions to this book of Maria Sági and Christopher Bache. These fellow researchers concur that we are moving through a major scientific, psychological, social—and hence spiritual—shift as a species.

There is little training available for such profound shifts underway in human perceptions and consciousness. Such a transition may indeed be unprecedented in the history of our species. From the mouths of caves to the starry dynamo of the intelligent cosmos, it is a wondrous journey that lies ahead. New horizons are unfolding as we speak (or read). As a human family, we are bound to embark on this journey together.

This chapter, and this book, aim to show the reader that we live in a participatory reality. Our universe is an interconnected Akashic Field. The Akashic reality we happen to be a part of does not wish for humanity to stand silent on the sidelines. We may only need to call out and say, *We are here!*

HEALING IN THE AKASHIC FIELD

Maria Sági

Healing in the space-and-time-transcending form is one of the most exciting proofs of the reality of the universal information field described by Ervin Laszlo. Phenomena of "spiritual" or "faith" healing have been known since earliest times, and there were important attempts to understand these phenomena in every culture, in every part of the world. But the mainstream paradigm of modern science did not offer a favorable environment for these attempts. Until recently, that paradigm could not find a place for mind and consciousness in its world picture and could not account for how physical healing could take place through mental—or even purely informational—means.

In the modern world, remote healing has been relegated to the realm of fiction. This is now changing. Compelling evidence has been amassed in experimental parapsychology and in the quantum sciences that the phenomena of information-based healing are real and important. The

Akashic Field theory developed by Laszlo offers an explanation of these phenomena and therefore represents an important development in the contemporary healing arts.

Here I undertake to link the new paradigm pioneered and developed by Laszlo with the phenomena I encountered in my twenty-year-plus experience as a healer.

A new paradigm is entering the mainstream of science. Scientists realize that space is not empty, and that what they have been calling the quantum vacuum is in fact a cosmic plenum. It is a fundamental field validating the ancient concept of the Akasha. As Laszlo tells us, the Akashic Field interconnects all things and conserves the trace of all things. It is a constant and complete memory store of the universe. The existence of interconnection and memory in nature is no longer fantasy: it is an established fact. And it is directly relevant to the practice of healing.

Healing can be practiced not only locally, but also nonlocally. This is possible thanks to a direct connection between the mind and consciousness of the healer and that of the patient. This connection is mediated by a universal information field Laszlo calls the Akashic Field.

The living individual, as all quanta and systems of quanta in the world, is not above or beyond the Akashic Field, but is embedded in it. The morphic pattern of one individual—his or her informational trace—interacts with the morphic pattern of other individuals. Their interactions shape the cluster of vibrations that defines the species. I call this collective cluster "the generic pattern of the species." It is the norm of viable functioning for all members of that species.

The generic pattern of the species integrates the morphic

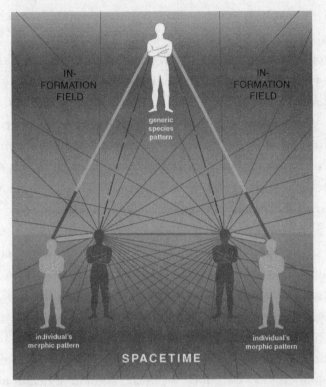

The figure above illustrates the hypothesis that all people, including the healer and the patient, are embedded in a cosmic information field that connects people and conserves the trace of their thinking and acting. Every living system has his/her cluster of waveform vibration constituted by information in the Akashic Field. The information that constitutes the cluster is real, and it is conserved. It is inretrievable in principle from the field.

pattern of the members of the species. In the generic pattern differences between individuals are canceled out. In the context of healing, the generic is the pattern of reference for health and vitality.

We are capable of maintaining our organism in a state

of health as long as our individual morphic pattern matches the generic pattern of the species. The information that determines the functioning of individual organisms is distributed throughout the atoms, molecules, tissues, cells, and organs of the body. Every deviation of the individual's morphic pattern from the generic pattern means a weakening of the individual's life energy and the diminishing of his or her level of health. If not corrected, it invites the onset of disease even before the symptoms of a disease appear in the body.

The match between the individual's morphic pattern and the generic pattern of the species is oriented by the subject's mind and consciousness. If the mind and consciousness of the subject is open and sensitive to alignment, the health-promoting match can be achieved. The information sent by the healer to the patient promotes and reinforces this match. When it is achieved, the patient's organism is able to maintain its vital functions in the domain of health and viability.

Malady and malfunction are errors in the way individuals match their individual morphic pattern to the generic species pattern in the Akashic Field. Life on the planet could emerge and can persist because a significant percentage of living systems could achieve the match of their individual morphic pattern with the pattern of their species.

In traditional societies people made intuitive and spontaneous use of the information they needed to create this match. Shamans, medicine men, and spiritual leaders above all in ancient Egypt, China, India, and Mexico have been promoting this match intuitively. Thereby they not only reinforced their own vitality, but contributed to a higher level of health in their tribe and society.

It is almost impossible to account for the rich variety of the remedies and healing traditions adopted by traditional cultures, but we can note that they all link the healing of the physical body with the healing of the spirit. Ancient cultures accepted the reality of subtle energies and used the bioenergetic processes that take place within and around the physical body in their healing. Although they go by different names, subtle energies and related spiritual elements have been basically the same in all cultures and all periods in history.

Healers now know that psychological attributes such as constructive beliefs, positive thinking, and healthy aspirations have an important role in redressing the maladies of the body. But Western medicine dismisses these psychological approaches and focuses almost entirely on the biochemical and physiological manifestations of disease. This predilection has historical origins. When the mechanistic/materialistic paradigm of classical physics became dominant, the healing of the body and the healing of the mind went their separate ways. The observation and therapy of the body became limited to the domains of the natural sciences, and the functioning of mind and consciousness were assigned to religion. Nonlocal-information-based healing was relegated to the domain of religion and spirituality. Today, this separation of science and spirituality is no longer accepted. Some age-old seemingly spiritual insights have been rediscovered as bona fide elements in science.

In my practice, I use information as the instrument by which to effect healing. Since all organisms are basically information-based entities—and are imbued by the active form of information Bohm called in-formation—through in-formational means we can treat the cause of an organic malfunction, and not only its symptoms. Given that the information that imbues the myriad processes of the organism is nonlocal (intrinsically interconnected and conserved), we can treat the condition that is the cause of a disease even if it occurred in the past. This contrasts with the mainstream method of Western medicine where (except for branches such as neuro-psychoneuroimmunology) diagnosis and treatment are largely limited to the actual physical-physiological condition of the patient.

UPGRADING THE UNOBSERVABLE DOMAIN OF INTUITIVE HEALING TO THE OBSERVABLE DOMAIN OF SCIENCE

Yogis and spiritual leaders tell us that we live in a world of polarities and complementary opposites. For effective healing, we need to take into account these polarities and opposites.

As coherent interactive elements in an interactive universe, we are not only able to perceive and respond to the effects reaching us from the outside world, we can also respond to and act on the world; we are in constant communication with it. Coherence comes about through an open and mutually responsive interconnection among the entities that make up the universe. Complete coherence, known as *supercoherence*, indicates the coupling of our intrinsic interconnections with the system of our wider—social and ecological—environment. How can we perceive the interconnection between our organism and the systems that make up our vital environment? It appears that we perceive the empirically unobservable information that reaches us from our environment through a channel of intuition sometimes called the sixth sense. Our perceptual system extends beyond our body and its five senses.

The wider and deeper scope of our perceptions is evident in all domains of our experience. Much as we sense changes in the weather, we spontaneously perceive the various and often changing qualities of the energies that affect us. Our body responds to vibrations that our sensory organs cannot perceive. Depending on these perceptions, we become energized and activated. Or we become ill. The more we pay attention to these beyond-sensory intuitions, the more we are able to maintain our state of coherence in the world.

For ancient civilizations as well as for indigenous people living in nature, it was evident that we are an organic part of our environment, connected with it through constant interaction and intercommunication. These peoples did not suppress and ignore the subtle information reaching them in the form of spontaneous and typically extrasensory perceptions. The perception of what we now call subtle energies was part of their experience. For them, it was evident that nature did not consist of physical matter alone—that the material body is only one dimension of the living organism. The energy and spirit that permeates the world was an effective factor guiding their life.

Hermes Trismegistus wrote, "As above, so below." We are a part of the ecology of the earth, and beyond that of the universe. Our forefathers knew this and they lived with subtle intuitions both on the natural and the supernatural level. They took account of spiritual, emotional, and energetic dimensions and systematized their knowledge of the extrasensory dimension as well as their knowledge of the body. They could heal in the spiritual dimensions as much as they could in the material dimension. They sought to recover the wholeness and health of the body through the spiritual dimensions associated with the body.

Ancient cultures gave various names to the beyond-sensory domain that carries and transmits information in nature. In Japan, Shintoism called it *ki;* in Chinese Taoism it was called *chi*. Ancient Hindu philosophy named it *prana,* and the traditional Judaism of the Near East knew it as *chaim*. In Europe, Pythagoras and the ancient Greeks regarded it as the fire at the center, while the alchemists of the Middle Age spoke of it as *azoth*.

The aboriginal peoples of Australia as well as some nature tribes in Africa and Oceania still intuitively know and follow subtle energies and cooperate with them. They commune with the earth and its inhabitants, and they respect nature and the order of nature. They are able to detect earth energies, and they know where to build their abodes and which places they should avoid. They are "psychic" in the sense that they are able to perceive and transmit information spiritually. They know, for example, if a family member left at home becomes ill or is facing a threat. They are able to assess the behavior of animals when they sense danger. They observe and account for minute changes in their environment and derive information from them and conduct their lives accordingly. They are acquainted with the flora and fauna of nature and know what plant or animal products cause harm to the body. Their spiritual guides can remedy not only the ills of the body, but also of the spirit.

The deep history of medicine is the story of how the unobservable spiritual world of classical healing became part of the experience-and-experiment-based world of modern medicine.

Some cultural precedents of remote healing

Before I introduce my remote-healing method, I wish to give a short account of the methods on which that method is based. The methods I discuss reflect the wisdom of cultures that have evolved and been perfected through thousands of years. We do not know how old they are, but findings—such as scrolls and instruments—testify that they have been in existence well over five thousand years. It is noteworthy that modern science can bear out the validity of many of these ancient systems of healing. Current experience-and-experiment-based techniques in modern medicine provide an insight into the way they function.

The first method I wish to discuss is acupuncture as it appears in TCM (traditional Chinese medicine). The mode of operation of this healing is to insert thin acupuncture needles at specific points in the body. By this method the healer reestablishes healthy information-and-energy flows in the body. The difference between this traditional method and the method I use is that I rely for information transmission on the universal information-and-memory field known as the Akashic Field. Following Erich Körbler, to effect healing, instead of needles I use lines and symbols at the body's acupuncture points. These can also be applied at a distance.

A basic system of symbols is used by healers around the world. The symbols rediscovered by Körbler are similar to the generally used systems. In ancient times different cultures used the same kind of symbols, even if they lived thousands of miles apart.

The classical Indian wisdom culture is another pillar of my method. I developed Indian wisdom–based chakra diagnostics and therapy to apply to cases in which the health problems of the patients derive from events in their past. The causes of their problems can be identified through this method, and the harmful information can be neutralized. The objective is to synchronize the two hemispheres of the brain. The synchronized state appears to open the way for retrieving the information that codes the normal, healthy state of our species.

I now move to healing methods evolved in ancient China. The manifestation of cosmic energy is a part of yin-yang theory in China's Taoist culture as well as in traditional Chinese natural philosophy. These systems recognized that any part of a system or ensemble can only be understood and effectively treated in its relation to the whole.

Yin-yang theory is based on the unity of two polar opposites and is a useful way of representing processes of change in nature. This system of thought is present in Chinese medicine, according to which the harmony between two poles, the yin (the passive, receptive, lunar) and the yang (the active, creative, solar) is achieved through the flow of *chi*, the fundamental life force. Its main method is acupuncture. Its objective is to restore the balance of energy flow within the body.

During the past four decades, acupuncture became one of the most widely analyzed fields in medical research. Tens of thousands of research projects have been carried out to measure the physical carriers of energy flow in the body. They offer experimental data regarding the position of the acupunctural points in the body. It was known that the stimulation of peripheral acupunctural points causes the activation or deactivation of specific parts of the brain. Experiments by Niboyet, Eőry, and Frenyó have now shown that oxygen consumption, carbon dioxide production, and skin resistance are measurably different at acupuncture points from those at other points in the body.[17]

We should remember that in the functioning of the immune system electromagnetic waves are produced with a dominant frequency of 633 nm (nanometers), although other frequencies also activate a response. (The wavelength 633 nm is that of the color cherry red. It impels cells to reach a higher level of coherence.) In the case of disease, the sick organ or organ system can be activated with coherent radiation of the 633 nm wavelength. Radiations at this wavelength not only produce a local salutary effect, but regulate and stimulate the entire immune system. Applying healing symbols at these acupuncture points can restore functional balance in the entire malfunctioning organism.[18]

The physical reality of the meridian system has been experimentally demonstrated by the increased emission of biophotons along the meridian lines. This was reported by famed biophoton researcher Fritz-Albert Popp (personal communication). My own experience confirms that the meridian system provides a reliable basis for diagnosis as well as for healing, both proximally and remotely.

LINES AND SYMBOLS IN THE ANCIENT HEALING ARTS AND IN TODAY'S NEW HOMEOPATHY

In the 1980s, Austrian researcher Erich Körbler developed the method of New Homeopathy, a healing system based on Chinese acupuncture. Körbler applied to the acupuncture points lines and combinations of lines rather than pricks by needles. He found that in the extremely high frequency band of the electromagnetic field, even small differences in the value of conductivity become decisive. A graphite pencil line produces a specific level of conductivity when drawn on paper and drawn on the skin. It functions as an antenna, absorbing and emitting waves of a particular frequency.[19]

Körbler discovered a systemic principle for analyzing the distribution of fields and their polarity. He showed that in the domain of very high frequencies, geometrical symbols act as connecting elements. With their help we can manipulate the content of the information reaching the body. In this way we can strengthen, weaken, or alter the effect of electromagnetic radiations on the body. Some lines and line combinations function as "antennas" in the patient's EM field. Each such line or combination of lines possesses a specific code for changing or restoring energy balances in the organism. They affect the body and can correct flawed information, thereby producing healthier energy states. I developed the healing system based on these principles, known as the New Homeopathy. This enables me to successfully practice both proximal and remote healing in hundreds of cases.

Consistent with Körbler's finding, anthropologists found that some ancient cultures drew parallel lines and isosceles crosses on the body of a sick person, using these for healing in addition to medicinal herbs. Many of these cultures, although located far from each other, used similar lines and symbols for their healing.

These remarkable synchronicities must have had proven beneficial effects. There are relics all over the world—including in Indonesia, Polynesia, Sudan, Nigeria, Brazil, and North America—that show that parallel lines and other symbols were used by shamans and medicine men. These findings received significant confirmation when in the autumn of 1991 an Austrian tourist found a corpse on the Italian side of the Ötztal Alps, about sixty kilometers from the Austrian-Italian border. Following detailed testing, the frozen corpse was shown to be about fifty-three hundred years old. The melting of the deep cover of ice revealed a brown-skinned man with parallel lines over his back, legs, and wrists, and an isosceles cross on his knees. They found a stone knife, a bronze ax, and a bow with arrows next to him. Autopsy revealed that the subject suffered from an intestinal disease that corresponds to the tattooed parallel lines on the corpse. The position of the symbols on the body indicated the nature of the disease suffered by this fifty-three-hundred-year-old man. Fifteen groups of blue-black tattoos were on his body, painted with coal dust. The forty-seven parallel lines organized into these groups were not drawn at random; remarkably, they show a definite correlation with the system of ancient Chinese acupuncture. In one form or another, healing by lines and symbols was present in the great spiritual traditions of the world.

The system of thought supporting this practice included the worldview of the Sanskrit culture's concept of Akasha, where Akasha denoted the fifth and deepest dimension of the world beyond the four dimensions of earth, air, fire, and water. *Akasha* was sometimes used in the sense of "sky" or "ether" and came to be seen as an ethereal field underlying the observed world.

The concept of the Akasha was present in the Vedic texts of India as early as 5,000 BCE. In the Vedas its function was identified with *shabda*, the first vibration, the first ripple that makes up our universe, and also with *spanda*, described as "vibration/movement of consciousness." The contemporary Indian scholar I. K. Taimni wrote:

There is . . . a mysterious integrated state of vibration from which all possible kinds of vibrations can be derived by a process of differentiation. That is called *N.da* in Sanskrit. It is a vibration in a medium . . . which may be translated as "space" in English. But . . . it is not mere empty space but space which, though apparently empty, contains within itself an infinite amount of potential energy. The relevance of the Akasha is due to its comprising the concept of "spirit."

This formerly esoteric classical notion is sustained and substantiated in science. In quantum physics, observations and calculations reveal that at the ultrasmall dimension, space is not empty and smooth. It is "grainy," filled with waves and vibrations. When physicists descend to the ultrasmall dimension, they do not find anything that could be called matter. What they find are waves and clusters of standing or propagating vibrations.

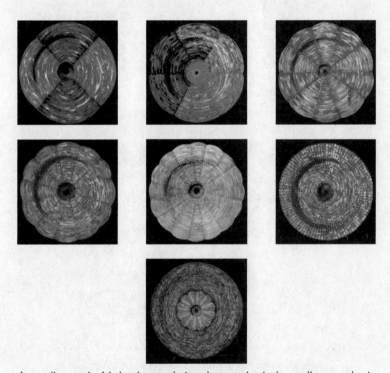

According to the Vedas, human beings have a physical as well as an etheric body. The layers of the etheric body are interconnected. Cosmic energy flows through seven chakras and more than three hundred and fifty thousand "nadins" (energy channels). The medicinal application of this classical concept is Ayurvedic medicine, based on the energy centers known as chakras.

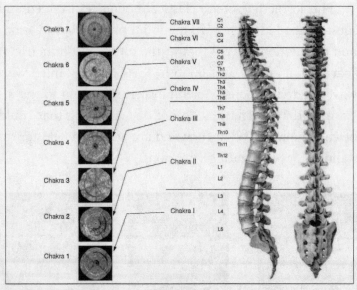

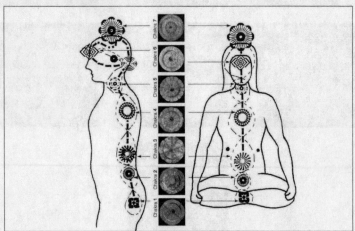

The Relationship Between the Chakras and the Vertebrae

The chakra pictures show the relationship between the chakras and the energy system of the body.

first chakra, the sacral vertebrae and lumbar 5, 4, 3
second chakra, the lumbar 2, 1, and thoracic 12, 11
third chakra, the thoracic vertebrae 10, 9, 8, 7
fourth chakra, the thoracic vertebrae 6, 5, 4, 3
fifth chakra, the thoracic 2, 1, and cervical 7, 6, 5
sixth chakra, the cervical vertebrae 4, 3
seventh chakra, the cervical 2, 1

Ayurveda, traditionally called "the science of life," proves to be an effective holistic healing method, practiced for more than three thousand years. It maintains harmony between body, mind, and spirit, and thereby it can reduce vulnerability to disease. Its effectiveness is confirmed by modern medical research. The harmonization of the chakras plays a definite role in the maintenance of health and the prevention and cure of disease. Some branches of medical science adopted as their goal the investigation of the role of chakras and other fine energies in the body.

Italian brain and consciousness researcher Nitamo Montecucco developed a method to test the level of synchronization of the subject's left and right hemispheres. He connected a computer with an EEG device and analyzed the level of synchronization with the help of a specially designed computer program. He called this synchronization-measuring device the brain holotester.

One of his experiments measured the EEG effect of the Sági chakra therapy.

In these experiments the healer's and the test subject's EEG waves are measured at the same time. In a remarkable instance, at the beginning of the experiment, the healer's synchronization level of his left and right hemispheres rose to 97 percent, a level that normally occurs only in deep meditation. The patient's synchronization level of the left and right hemispheres was 35 percent, and this corresponds to the alert brain state. During the experiment, the patient looked at each of the seven chakra maps in turn, while the healer held his left palm above the patient's right hemisphere. He held the bioindicator in his right hand, testing the activity of the relevant chakra. After a few minutes, the patient's left and right hemispheres' synchronization increased to 72 percent, while the healer's continued at 97 percent. At the end of the experiment, the patient's level of synchronization first rose to 81 percent, and then, after about ten minutes, it reached 91 percent (the healer's level of synchronization was first 91 and then 95 percent). The high level of synchronization of the healer's and the patient's EEG demonstrated that chakra therapy produces the measurable physiological effects.

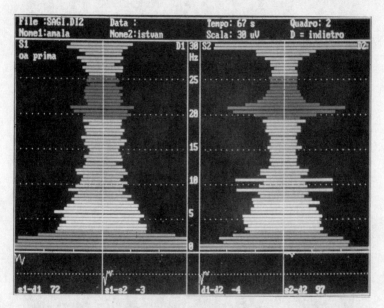

In an experiment, the EEG waves of the subject's brain and the healer's brain were measured to determine synchronization between the left and the right brain hemispheres. The left side of the plate shows the level of synchronization of the left and the right hemisphere of the subject, and the right side shows the synchronization level of the healer. S1 is the left hemisphere, and D1 is the right hemisphere of the subject. S2 and D2 represent the left and right hemisphere of the healer. Synchronization is shown by the equal length of the horizontal lines (representing the left and the right brain hemispheres) projecting from the vertical white line at the center. In the beginning of the experiment, the healer's brain hemispheres showed synchronized levels consistent with deep meditation, that is, 97 percent. The patient's synchronization level was 35 percent, corresponding to the alert brain state. After a few minutes of exposure to the Sági chakra therapy, the subject's level of synchronization increased to 72 percent, and the healer's remained at 97 percent.

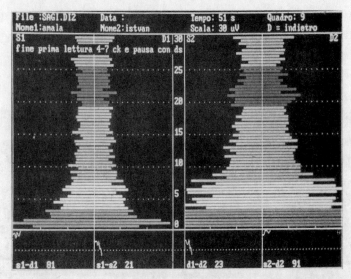

The subject's left and right hemispheres' synchronization level progressively increased to 81 percent, and the healer's level decreased to 91 percent.

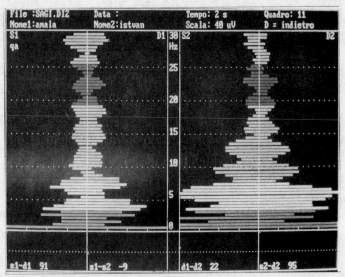

After ten minutes, the subject's level of synchronization reached 91 percent, and the healer's reached 95 percent.

PRACTICING SPACE-AND-TIME-TRANSCENDING HEALING

Healing with information is space-and-time independent. In light of my experience it is rendered possible by the "tuning" of the healer to the Akashic information and memory field, the field that conserves and conveys the morphic pattern of the individual as well as the generic pattern of the species. The healer selects the information that creates an improved match between the individual and the generic patterns, thus reinforcing the energy balances of the patient.

The process of tuning oneself to patterns in the field can be compared to tuning a radio, a TV, or an online computer to the electromagnetic field. We need an artificial device to transduce the signals in the frequency range that is decodable by our device. In the case of information-based healing, that device is the brain. Our brain picks up the signals that reach it from the patient as well as from the world at large. But it can do so only within a limited range. In the normal waking state this range is relatively narrow.

Because of the widespread belief that all the information that reaches our brain is conveyed by our bodily senses, most people maintain that their experience is limited to the domain of sense-perceivable information. However, now we know that we enter into an altered state of brain and consciousness, and thereby we can extend the range of our perception beyond the receptive range of our senses.

Receiving information from the world is a wide-band-frequency-reading process. The healer attunes to the frequency of the patient and focuses on the information she receives. She scans the informational trace of the patient's organism (the patient's morphic pattern) in view of finding abnormal features. These features are to be corrected by promoting the match between the patient's morphic pattern and the generic pattern of the species.

There are a variety of methods at disposition for selecting and fine-tuning the healer's receptivity. We can make use of any suitable device and modality for this purpose, such as a pendulum, a dowsing rod, direct touch, imaginative "channeling," and others. We can also designate a substance or an object as a "witness" and focus on it. The witness can be a photo of the patient, a sample of his or her writing, an audio or video recording, a lock of hair, or a sample of blood or saliva. Every sample of the patient's body facilitates the retrieval of information from the organism of the patient.

If patient and healer are emotionally connected—as members of the same family or as close friends—there is no need for a witness: the healer can tune in to the field and retrieve information on the patient's condition through that intermediary person. If healing takes place in a group and someone in the group knows the patient, it is nearly as easy to attune to information on the patient as if he were within the group.

In the next step, the healer evaluates and orders the information she gleaned by accessing the patient's individual morphic pattern. Experience shows that every healer has his or her own system for doing this; many modes of diagnosis and healing are available, and the difference between them is often a matter only of cultural preference. In the Western world, the handiest approach is to make use of the instruments of mainstream medicine, examining and testing the pertinent patient's organs and organ systems one by one. Homeopathic healers can proceed according to the classical homeopathic system developed by Hahnemann, and physicians working with the method of psionic medicine can complement mainstream and homeopathic treatments by using the radiesthetic method. Anthroposophical healers effect their diagnosis with the help of the anthroposophical system. The healer can also work on the basis of the major Eastern diagnostic systems, such as the Chinese, the Ayurvedic, and the Tibetan. Healers who can perceive the aura of the human body can perform their diagnosis on the basis of the appearance of the various auric layers. The color and density of the aura offers clues for their diagnosis. All these methods can help the healer to ascertain the particular points in the organism that require treatment, and to evaluate the information regarding the condition of the patient.

The information received by the healer indicates both the physical condition of the patient and the nature of his or her problem. This information is not limited to the time the examination is carried out: it is distance independent as well as time independent. A diagnosis from a remote location and from a later point in time may be as detailed and thorough as one based on direct examination. The healer can discover whether the health problem is temporary or chronic, and whether it originated recently or further back in the past. She can also determine the causes of the problem, and to what extent they are due to harmful environmental conditions (even of such relatively "soft" varieties as electrosmog and earth radiation). Then the healer needs to decide whether it is necessary to change the patient's milieu, or whether more favorable conditions can be created in the patient's current milieu.

The therapy to be recommended by the healer can be based on any of the available diagnostic systems. The healer can also prescribe a variety of remedies, including standard allopathic cures, diet, phytotherapy, homeopathic remedies, and others, and can advise the patient to procure the substances and create the conditions for the cure.

It is not necessary for the patient to be aware of the progress of the therapy. The important thing is that he or she invites and welcomes it. A negative approach of out-of-hand rejection, and even dogmatic skepticism, can block the transmission of healing information.

When the transmission is achieved, the entire organism of the patient is affected, not only the dysfunctional parts. The effectiveness of the healing can be measured by follow-up diagnosis, conducted either by conventional or by information-based methods.

In regard to the physics of healing, we should note that space-and-time independence suggests that besides electromagnetic waves, scalar and other, still more esoteric quantum waves transmit the information. This is likely because the healing information is not likely to be carried by electromagnetic waves alone. In that case the effects would attenuate with growing distance in space and in time. But there is evidence that this is not the case.

On the basis of my own experience, I am convinced that the information I receive and send is transmitted by the universal information-and-memory field we call the Akashic Field. The grounds for this conviction are evident. The pertinent information could not be transmitted directly from the organism of the patient because that information would be limited to the patient's present location and current condition. But I and other healers can receive information on the condition of the patient from any point in space and from any point in his or her life—even from just after birth, and sometimes before that, from the period of gestation in the mother's womb. I can concentrate on any period I wish and can find the point in time that is pertinent to the health problem of the patient. Often I can pinpoint the traumatic event that created the health problem in the childhood of the patient by interviewing the mother or a friend or family member who witnessed the problem-generating event. The pertinent point may even have preceded the current lifetime of the patient. Although most health issues have roots in one's own lifetime, some deep-rooted problems could have originated in a previous life.

I can get information about the patient's earlier condition and relate it to his or her current condition. Doing so is essential for effective diagnosis, as many and perhaps all organic ills and dysfunctions have their origins in the patient's past. That I can move back in his or her life history to any point after birth, sometimes to the period of gestation, and even to a prior existence, is meaningful indication that a nonlocal field is conserving and conveying the information I receive on the patient's condition.

HEALING PSYCHOSOMATIC MALADIES

Although there is widespread agreement that nearly all diseases have a psychological origin, finding the root cause of a disease is still difficult. This is because, among other things, people react differently to different events—there is no dependable connection between objective physiological conditions and subjective psychological responses. A substance or stimulus that makes one person sick could pass unnoticed by another.

Environmental stress and the organism's reaction to it can produce a wide range of maladies. It can cause and sustain chronic or frequently recurring diseases, such as high blood pressure, heart attack, rheumatoid arthritis, chronic diseases of the respiratory tract or of the intestinal system, and others.

There are also more subtle causative factors. Everyone knows the stress resulting from emotional friction in everyday life. This occurs frequently in the close emotional bonds that come about within the family, but it also happens in less close relations with colleagues and friends. Even a slight misunderstanding or a misinterpreted remark or opinion can leave a profound trace. Both real and imaginary ills weigh heavily on the equilibrium of our emotional life. They create a burden also for our physical well-being and health. We know that physical fatigue goes hand in hand with a disheartened condition, and that physical diseases can be catalyzed by grudges one has held for a longer time.

Erich Körbler emphasized that we react to events both from the inside and from the outside of the organism, and our autonomous nervous system is unable to separate these impulses. Intellectual or emotional experiences can cause stress inside the organism and are as likely to disturb equilibriums within as disturbances originating on the outside. Over the longer term, both kinds of stresses can provoke disease.

Psychosomatic problems have to do with the content of the information that affects the individual. If the content is harmful, this appears as an acute psychological and physical malfunction. Then it is emotionally disturbing information that manifests as a physical symptom. Psychosomatic symptoms can readily be identified as they keep reappearing in spite of physical treatment. Information about old, long-forgotten traumas is conserved in nature's information field and continues to act not just as an element in the psychological makeup of the person, but as a causative agent behind a physical malfunction.

The human organism is a complex network in which the operation of the parts is conditioned by the coherence of the whole. The whole consists of a great number of hierarchically organized systems (organs, cells, molecules, atoms, particles) that are themselves constituted of a multitude of subsystems. None of these components can change without triggering a change in the others, and hence in the system as a whole.

When we explore processes within the organism, we find that information travels more rapidly throughout the system than standard biophysical explanations can account for. The organism is a macroscopic quantum system, governed by interconnected spatial, temporal, and energy hierarchies, linked by quantum energy flows. The highest level among these hierarchies is the mind or consciousness.

The information content of the functions of the mind, such as emotions, intuitions, and thoughts, affects all parts of the organism, whether one is conscious of them or not. In the brain, every impact is integrated with the rest and produces either a favorable or an unfavorable effect on the organism. Every cell constantly informs the brain on its condition. Evaluation in the network is nonlinear, oriented toward creating or safeguarding optimal conditions for the functioning of the organism.

Testing the coherence of the organism

In the human body there are various forms and manifestations of coherence. Of particular importance is the organization of the energy flow. Its dominant direction is from the top of the head to the tip of the toes. The polarity of the cells and cellular systems is aligned to produce this flow, and this ensures communication between the parts and the organs crucial for the body's functioning.

If the polarization of a part of the system is reversed, the coherence of the flow is broken and an energy blockage results. With the methods of new information-based medicine we can detect the disturbance in the energy flow of the organism even before it provokes physical symptoms. In an energy blockage the affected organ or body part is isolated from the coherent communication in the rest of the body; a disturbance occurs in the flow of information between the affected part and the central nervous system. To detect energy blockages is an important task of diagnosis in information-based medicine.

If we want to preserve the healthy functioning of the organism, we need to sustain the coherence of the body. The self-healing potentials of the body cannot operate unless this coherence is present. However, there are also disadvantages to the coherent operation of the organism. If there is a local lesion (an injury) in the body, owing to the coherence of the affected part with the rest, this will affect the entire organism, and this can produce a diminished sense of well-being in the whole organism. The local problem is amplified by the coherence of the body's information and communication system.

Information-based medicine explores how energetic influences and information processes outside the organism affect the overall coherence of information in the body. It maps these connections regarding the extent to which adaptive resonance is created between the body and the environing world. Disease is a disturbance of the resonance between the environing world and the body. The highest level of adaptive resonance means health.

In my method of healing, the nature of the energetic condition of the body is shown by the movement of the bioindicator. Changes in the information affecting the various parts of the body produce changes in the energy flow. We need to identify information that is harmful to the body and attempt to neutralize it. We can transform harmful inputs into beneficial ones and thereby improve the capacity of the organism to make use of coherent energy flows.

The functionality of the vital processes of the organism is determined by the interaction of the body with the embedding electromagnetic field. This has been experimentally shown by Nobel-laureate brain researcher Roger Sperry. Sperry demonstrated that the consciousness associated with the neocortex displays the functioning of the entire nervous system. Consciousness reflects the functioning of the brain and the neuronal networks. Positive or negative contents of consciousness affect all the cells of the body. Positive contents of consciousness promote the coherent functioning of the system. For good health, every cell in the system needs to function in harmony with every other cell. Any stimulus, whether external or internal, that proves positive promotes the coherence—the harmonious functioning—of the system.

Sperry's discovery has been confirmed by Bruce Lipton's experiments on endothelial cells. Lipton demonstrated that on the cellular level instructions arriving from the mind can overwrite instructions from the organism. The experimental finding is that endothelial cells in a tissue culture change their behavior in line with signals from their environment. If the environment is rich in food, they move rapidly toward the source of nourishment. In a toxic environment, the cells distance themselves from the source of the harmful stimuli. But Lipton found that signals coming from the central nervous system have priority over signals of local origin.

Körbler found that positive contents of consciousness bring an energy surplus to the organism, while negative contents depress the energy levels. With favorable inputs, cell voltage builds up to 60 mV and even higher. The molecular structure of the cell develops so as to enable further connections. Because of the coherence of this process, effects spread over the entire organism.

In my experience, the bioindicator reacts sensitively to changes in the psychosomatic condition of my patient: it indicates whether a subtle psychological factor, such as a thought or idea, is tolerable or intolerable for him or for her. It also shows how psychological changes act on the physical state of the body. Negative, destructive, or harmful events affecting the mind reappear on the physical level and affect bodily function. This holds true even if the harmful event concerns another person, or if we merely see it on-screen. All survival-threatening acts trigger one of the standard responses: flee, fight, or freeze.

Using the psychomeridian in the Akashic Field

Psychosomatic problems prove difficult to treat by conventional methods. In the long term, bringing them into conscious awareness or applying medication does not help because previous conflicts may consciously or unconsciously be reactivated and cause problems. In some cases, a traumatic experience can go back to the period right after the subject's birth. For example, when a newborn fails to get the mother's milk (perhaps due to a problem of the physiology of the mother) and is fed with a synthetic formula that contains milk of animal origin, this constitutes a traumatic experience. As a result the infant may develop a lasting malady, such as neurodermatitis—an allergic response to milk of animal origin.

As a rule, a traumatic experience, regardless of when it occurs in life, produces stressful information, and this shows up in the body. The negative information produced by a traumatic experience needs to be changed—it does not vanish by itself. Past experiences influence the functioning of cells in the present. Thus, if we wish to eliminate the effect of a conflict that has negatively affected the organism, we need to transform the stress-provoking information. This we can do thanks to the discovery of the psychomeridian.

The psychomeridian explained

For the exploration of the psychological causes of a malady, a number of methods are offered in psychiatry and psychology, as well as in some Eastern healing systems. The procedures suggested by these approaches tend to be lengthy and costly. However, there is now a better option: the so-called psychomeridian, discovered by Körbler. This offers a simple and reliable way of identifying the psychological causes of an illness. Usually a single session suffices for discovering the time of the illness-provoking event, and opening the way to discovering the nature of that event.

In a complex quantum system, every part represents and actually includes the whole. This is true of the human organism. Our body has evolved a system of points that ensures that the most important information can "read out" on the body's surface. The various reflex zones of the body (the soles of the feet, the palms of the hand, and the face and ears), as well as the meridian system, represent energy projections that map the entire organism.

With the psychomeridian, Körbler identified an acupuncture-type point that enables the healer to discover the timing of the traumatic events that may have occurred in the life of an individual. The testing identifies the traumatic events experienced by the patient, as well as the degree of trauma produced by these events. The psychomeridian is a vertical line at the back in the middle of the head, starting from the atlas and ending at the top of the head. At this point of self-iteration, the body is capable of manifesting information on all the events in the life of that individual. At the atlas we can test events at the birth of the individual, while on the top of the head we can test events that occurred at recent times. The line can be subdivided and produces the chronological sequence of events in the lifetime of the individual.

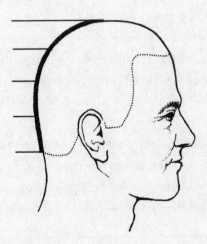

In testing the psychomeridian, we use practically the same procedure as using the other diagnostic points, with the difference that we conduct the testing at a different psychological level. These tests take us into deep waters: we address profound layers of the personality thanks to the connectivity and memory of the Akashic Field.

When testing the psychomeridian, we reach the points at which a trauma occurred. It often happens, however, that the subject cannot recall experiencing a traumatic event at that time. In that case, we, through the testing, are reviving unpleasant memories, and this can be a painful experience. The purpose of the therapy is to transform and if possible neutralize these unpleasant memories.

We start the therapy at a point in time when intense negative reactions become evident. We choose a symbol that calms the patient (visually perceived symbols can modify negative information). Once we have the appropriate symbol, we ask the patient to hold it in his or her left hand and look at it for a few minutes. Three or four minutes are usually enough to restore emotional balance. When the subject is composed again, the bioindicator responds with a powerful horizontal movement. Then we know that we have chosen the right symbol for altering the information produced by the trauma.[20]

The impact of therapy and additional effects

Experience shows that following the therapy, patients undergo significant changes in their psychological and physical makeup. They usually overcome their complaints

quickly and easily. If they suffer from depression and pho-
bias, they might not even require further treatment.

The therapy may induce the recollection of shattering
moments in the life of a person. He or she might cry or feel
physically sick. In an elderly male patient, the recollection
of a critical point in his infancy provoked crying like that of
a baby, accompanied by the spasmodic movements of arms
and legs typical of infants. But even if the recollection is not
conscious, the body reacts to the events as if they happen
all over again.

The neutralizing of the negative effects of a traumatic
event often creates positive change in the effects of other
traumatic events. The effects of traumas become neutralized. Such changes have been observed regarding traumas
occurring at early moments in life. The information of later
events is modified, and their negativity disappears from the
psychological makeup of the person.

In the case of allergies, patients often get rid of complaints that had prevailed for many years. In the background of long-standing psychosomatic complaints (e.g.,
neurodermatitis, milk allergy, asthma), we usually find
traumatic events in infancy or early childhood. Experience shows that the symptoms associated with these
events disappear when the information responsible for
them is modified.

To sum up

My more than two decades of practicing remote healing
together with the experience of hundreds of other healers

offers convincing evidence for the existence of a time-and-space-transcending information field in nature. This field functions as the medium that conveys information between healer and patient. It enables the matching of the individual's morphic pattern with the generic pattern of the species.

The emergence of healing based on information, whether proximal or distant, has major implications not just for individual cases of healing, but for medicine as a whole. A new era may be opening for the science of medicine, complementing, although not replacing, the mainstream approaches with a new "soft" approach, based on information.

PSYCHEDELICS-MEDIATED AKASHIC EXPERIENCES

Christopher M. Bache

Ervin Laszlo has eloquently described the Akashic Field through which billions of galaxies are interconnected and function as an integrated whole. He has given us a vision of our universe floating in an invisible cosmos of infinite potential, matter dancing in the quantum field, galaxies in-formed by Akasha, a cosmic intelligence that forms and in-forms all things.

In addition to the evidence Laszlo marshals for this concept from the frontiers of science, I can testify on the basis of my own explorations that the Akasha is a reality that can be experienced under the right conditions. One of these conditions is the conscientious use of psyche-delics. There is a deep correspondence between certain experiences that emerge in psychedelic states of con-sciousness and Laszlo's description of the Akashic Field. In this chapter, I want to illustrate this correspondence by sharing five experiences from my personal psychedelic explorations.

Shortly after beginning my career as a university professor in 1978, I began what would become a twenty-year inquiry working with LSD. Between 1979 and 1999, I did seventy-three therapeutically structured LSD sessions following protocols established by Stanislav Grof. After three low-dose sessions (two hundred micrograms), I did seventy high-dose sessions (five hundred to six hundred micrograms). In all of these sessions I was isolated from the outside world, cared for by a sitter, lying down with eyeshades and headphones, and listening to carefully selected music. A detailed account of each session was written within twenty-four hours.[21]

Over the course of this long self-experiment, I was taken through a repeating cycle of death and rebirth that propelled me into an ever-deepening communion with what I experienced to be the living intelligence of our universe. For the purposes of this essay, I'm going to assume that the Akashic Field either *is* this living intelligence in its entirety or is a *subset* of it. While I experienced this intelligence to be whole and seamless within itself, there were many layers to my initiation into it, as described in this passage from *LSD and the Mind of the Universe:*

As I have experienced it, consciousness is an
infinite ocean of experiential possibilities. When
we take these amplifying medicines, the mind
we drop into this ocean acts as a seed crystal
that catalyzes a certain set of experiences from

its infinite potential. As we are gradually healed, purified, and transformed by these encounters, the seed crystal of our mind is changed. In subsequent sessions, it catalyzes still deeper experiences from this ocean. If we repeat this process many times in a sustained fashion, a sequence of initiations into successively deeper levels of consciousness takes place, and a deepening visionary communion unfolds.

Because of the progressive nature of these initiations, I experienced different encounters with Akasha at different stages of my journey. The five experiences presented here do not capture the entire span of these encounters but are samples taken from selected points along the way that I hope will give a sense of the larger engagement. I've tried to choose experiences that are comprehensible on their own terms without requiring a great deal of background. If the number of samples was increased, the correspondence with Laszlo's theory would be even stronger.[22]

Laszlo writes, "The Akashic experience comes in many sizes, forms, and flavors, to all kinds of people, but all its varieties have a major fact in common: they convey in some form information on the world—on the world beyond the brain and the body." This is certainly true for the experiences described here. For the most part, I think the correspondence between these experiences and Laszlo's description of the Akashic Field is straightforward. It clearly demonstrates the presence of the Akashic laws of connectivity, memory, and coherent evolution. Therefore,

I have given more space to presenting the experiences themselves than to drawing out and explicating this correspondence.

FIVE CONTROLLED EXPERIMENTS

Before I begin, I want to mention that the access LSD gives us to these dimensions of consciousness is neither quick nor easy, but earned with hard work, as every serious researcher of consciousness knows but laypersons often fail to appreciate. Intimations of oneness may surface early in one's sessions, but to *stabilize* awareness at these deep levels requires undergoing intense cycles of purification, so that one's perception and cognition there can be clear and accurate. These episodes of "purification unto death" dissolve the boundaries of our being, not once but multiple times at successive levels of consciousness. Reaching beyond the boundaries of our *individual* life, they can even trigger episodes of *collective* purification that I believe impact the mind of our species directly. In what follows, I have softened but not eliminated this purification in order to focus on the communion that emerges from it.

Experiment #1. *Session 11*
Sessions 11 through 17 took me beyond linear time into a stable transtemporal condition that I came to call Deep Time. Deep Time is a dimension of consciousness in which the rules of linear time have been suspended, where one can experience different *spans of time* as being simultaneously present—past, present, and future existing concurrently

within some larger temporal horizon. At a deeper level, one can enter enormous tracts of time including even the distant future. Deep Time is not transcending time altogether; that is a different experience. The experience of Deep Time is a shift into a different order of time, a different way of being conscious with respect to time.

Laszlo has underscored the conserving function of the Akashic Field, that everything that happens in the universe is preserved in Akasha. The following session opens a further possibility, namely, that the Akashic Field also knows the (probable?) future, or that it in some essential way is not constrained by linear time. When one experiences these different temporal envelopes in a psychedelic session, it becomes compellingly clear that the universe has different temporal modalities built into it. As a result of many such experiences, I have come to believe that there are many layers to the tissue of time in the cosmos. I don't pretend to understand how this works. I only know on the basis of repeated experience that linear time is how time behaves inside space-time. As one moves to the "edges" of space-time consciousness, the rules of time change. The experience of *personal* Deep Time reported here was the forerunner to excursions into *collective* Deep Time that took place later on the journey. From this point on, time became more porous in my sessions.

S 11—Entering Deep Time[23]

The electrical spasms were intense, shooting me across the mattress. The music pulled on my darker spaces. My psychological anguish grew until I found myself trapped in a musical chamber of horrors. The tension was unbearable.

The music changed to tribal ritual chanting—complex breathing rhythms combined with shouts, grunts, and retching. As I let go to these sounds, I felt myself enter a primitive domain completely beyond any modern frame of reference. All familiar associations were suspended. Around me, through me, swirled fearful negative energies, elemental and barbaric. I was floating in a surging field of negative forces. I slowly became less frightened as there was less and less of "me" present to react to the experiences. As I dissolved into this field, I was emptied of all personal associations, hollow to anything but these ancient sounds, lost in another world, another time.

As the fierce cleansing receded and Beethoven's Fifth Symphony began its majestic entrance, a vision of my life began to unfold. It was so vast in scope and depth that I could not have taken it in had I not been so thoroughly disassembled by the prior cleansing.

The initial vision was of deep space, of galaxies spinning in silent rotation, countless stars suspended in sweeping galactic arms—portraits of the different magnitudes of time the universe operates by. Moving toward them, I began to dilate, to become larger and more expansive. I was being absorbed into what I was approaching. This was an extraordinary sensation. It completely changed my mode of experience.

Today it was as though I was young and old at once. I was my entire life experience with time collapsed, as though someone had turned my life on end and looked down its length, like looking through a paper-towel tube. Seen end to end, time disappeared and my soul appeared—the being beneath the stages of my life. Enduring relationships stood out in bold relief. Bonds with people, with ideas, with life tasks. People I had found and who had found me. Ideas that have circulated round me all my life, returning in different forms again and again. I was my past, present, and future—one—in a way that was certain to me, doubt not being possible.

Even more difficult to convey than the riddle of time is the extreme saturation of detail in this mode of experience, the extraordinary richness of layers upon layers of information. These details coalesced to form a "deep reading" of my life. The authority of this reading was incontestable; its truth was obvious. It was as though all the evidence was present. This way of knowing was unlike anything I had ever experienced before. It was a knowing that was not linear but whole. Not a conclusion but a seeing-things-in-their-entirety. It was a depth of vision so much richer than ordinary vision that to call it a "vision" lies. It was more a "being tuned" than a seeing. It was saturated textures of experience from different periods of my life symphonically orchestrated.

In these experiences, I not only saw and observed my future life, I tasted it experientially. I became my older self and felt its texture. It was not being a specific age or being older in some vague, generalized sense. I was experiencing the distilled residue of my entire life experience condensed into Now. With another person, it was to be aware of and appreciate your full history together. It was to see your lives happening and having happened simultaneously. With myself, it was to feel my life's currents, to know its essential themes. It was experiencing the larger flow of my life with time deleted and historical settings collapsed into my "whole person."

Experiment #2. *Session 19*

After the seventeenth session, I stopped my psychedelic work for six years. When I resumed it in 1990, my sessions took me far beyond my personal reality into a series of initiations into the universe and the Creative Intelligence behind the universe. This Intelligence never took a physical form. It was more a felt presence than something seen, but sometimes it manifested as a dynamic and articulate presence, as it did in the following session. In this session the Consciousness I encountered took me first into a field of intense collective suffering and then on an extraordinary tour of our universe.[24]

S 19—the Cosmic Tour

After a long opening, darker experiences began to emerge, but I was able to remain physically open and let them come. Again there was the frenzied, chaotic, physical, psychological anguish that I cannot put into words. At several places, I found myself wondering what all this pain was about. I was open and letting it come through me, but where was it coming from? I could not tell, but it broadened and deepened for a long time.

My consciousness was expanding and opening to more and more suffering. Eventually, I had the sense that my being stretched from horizon to horizon as I experienced a suffering that involved tens of thousands of people. Because I had gone into this state more slowly this time, it was not as confusing as before. Then I caught a glimpse of something behind the field of suffering, something enormous and familiar from the previous session. I reached more deeply into the suffering and eventually broke through to this larger dimension.

In most of my sessions, I have simply been carried along from one transpersonal experience to another. Only once before have I had the experience of being able to consciously direct my experience. Now it was happening again. A circle opened around me and created a space that became an arena of dialogue between myself and a larger Consciousness. I discovered much to my surprise that this field was responsive to my thoughts. When I first discovered this, I had the ecstatic sensation of confronting an enormous Intelligence that included and surrounded my own. "That's right," it communicated to me. "That's exactly what is happening."

I began to ask it questions, and it answered by orchestrating my experience in the circle. It was an extremely subtle process, and the line between "my" consciousness and this larger Consciousness was often invisible to me. At times my reaction to an answer interacted with what I was being shown to sidetrack the lesson being given. I learned that I could stop these unwanted deviations by taking control of my thoughts. I could "clear the board" by stopping my reactions and waiting for the space I was in to clear. Once my mind was still, the lesson would continue.

After some intervening experiences, I was brought to an encounter with a unified field underlying all physical existence. I was confronting an enormous field of blindingly bright, incredibly powerful energy. This energy was the single energy that composed all existence. All things that existed were but varied aspects of its comprehensive existence. Experiencing it was extremely intense and carried with it a sense of ultimate encounter.

The experience then changed into a moving experience of a Cosmic Tree. The energy became a massive tree of radiant energy suspended in space. Seeming larger than the largest galaxy, it consisted entirely of light. The core of the tree was lost in the brilliant display, but limbs and leaves were visible around its edges. I experienced myself as one of its leaves. The lives of my family and close friends were leaves clustered near me on a small branch. All our distinguishing characteristics, what made us the individuals we were, appeared from this perspective to be quite minor, almost arbitrary variations of this fundamental energy.

I was taken around the tree and shown how easy it was to move from one person's experience to another's, and indeed it was ridiculously easy. Different lives around the globe were simply different experiences the tree was having. Choice governed all experience. Different beings who were all part of Being Itself had simply chosen these manifold experiences.

At this point, I WAS THE TREE. Not that I was having the full range of its experience, but I knew myself to be this single, encompassing Consciousness. I knew that Its identity was my true identity. Though I had taken monism to heart years before, I was now actually experiencing the seamless flow of consciousness into crystallizations of embodiment. I was experiencing how consciousness manifests itself in separate forms while remaining unified. "So that's how it works," I said to myself. The freedom was sheer bliss.

As I left the experience of the Cosmic Tree, the sensation of intense energy subsided, and I found myself to be once again in conscious communication with this vast, surrounding Consciousness. My experiential field was extremely clear.

For the next several hours, this Consciousness took me on an extraordinary tour of the universe. It was as though It wanted to show me Its work. It appeared to be the creator of our physical universe. It would take me somewhere or open me to some experience, and I would come to understand some aspect of the workings of the universe. Over and over again, I was overwhelmed at the magnitude, the subtlety, and the intelligence of what I was witnessing.

"That's incredible."

"I'm beginning to understand."

I was repeatedly left breathless by the beauty of the design I was seeing.

Sometimes I was so staggered by what I was seeing that I would stop, and It had to come back for me. "Keep up! Keep up!" It said, taking delight in my awe. Sometimes I was not sure what I was see-ing, and It would do something and suddenly everything would be-come larger and I would understand. Then It would take me on to something else.

This tour was the most extraordinary journey of my life. The vistas of intelligence repeatedly swept me into cognitive ecstasy. The irony, however, is that except for the small pieces I shall describe below, I am unable to re-create the details of what I saw. I simply don't have enough Ph.D.'s to fit the knowing I had there into my small Earth-bound mind. This does not lead me to question or doubt my experience. Even though I have lost large sections of the experience, I retain an unshak-able epistemological certainty that this knowing was of a higher order of knowing than any I am capable of in my ordinary consciousness.[25]

At one point, I was taken through a complex labyrinth of churning forces until I emerged above the turbulence into a wonderfully spacious and calm experiential field. I was told that we had come through the emotions of human experience. They had a restless, gnawing quality to them and composed such a mass of tangled energy that I was not surprised that they could blot out this subtler domain of peace and tranquility.

My elevation into this field felt like remembering, as did all my experiences on this tour. I was reawakening to levels of reality that I had previously known but had forgotten. Over and over again, I was reawakening to a level of experience I had left behind long ago. Remembering. It was not about "dying" at all but waking up and remembering.

I was then lifted into another "higher" and "larger" experiential field and then another. With each transition, I entered a new level of quiet and bliss-filled peace. It was as though an amnesia lasting billions of years was being lifted from me layer by layer. The more I remembered, *the larger I became. Wave after wave of awakening was pushing back the edges of my being. To remember more was to become more.*

Finally, I was lifted into a particularly spacious and peaceful dimension. As I remembered this dimension, I was overcome by an overwhelming sense of homecoming and felt fully the tragedy of having forgotten this dimension for so long. I cannot describe how poignant this was. Being fully restored to this dimension would be worth any cost. I asked what had happened, and It explained that we had left time. Then It said, "We never intended so many to get caught in time." It felt like time was simply one of the many creative experiments of the multidimensional universe I was being shown.

Though these experiences were extraordinary in their own right, the most poignant part of today's session was not the dimensions of the universe I was witnessing, but what my seeing them meant to the Creative Consciousness I was with. It seemed so pleased to have someone to show Its work to. I sensed that It had been waiting billions of years for embodied consciousness to evolve to the point where we could at last begin to see, to understand, and to appreciate what had been accomplished in our self-evolving universe. I felt the loneliness of this Intelligence, having created such a masterpiece and having no one to appreciate Its work, and I wept. I wept for its self-isolation and in awe of the profound love that had accepted this isolation as part of a larger plan. Behind creation lies a Love of extraordinary proportions. The Intelligence of the universe's design is matched by the depth of Love that inspired it.

Somewhere in here, I realized that I was not going to be able to bring the knowledge I had gathered on this journey back with me. The Intelligence I was with also knew this, making our few hours of contact all the more precious to It. There was nothing I was going to be able to do with this knowledge except experience it now. My greatest service was simply to appreciate what I was seeing. It seemed important to mirror existence back to its Creator in loving appreciation.

Experiment #3. *Session 21*

In the Cosmic Tour, I explored the universe as a reality outside myself and had accordingly experienced the Consciousness that orchestrated this tour as being "other" than me, despite the brief experience of nonduality in the Cosmic Tree. Two sessions later, I explored the universe as a dimension of my own being—not my personal being but the infinite Being I am part of. This shift reflects the broader transition that takes place as one moves from the subtle level of consciousness, where dualism is still intact, to the causal level of consciousness, where all boundaries are dissolved in Oneness.[26] The experience of Oneness in this session resonates deeply with Laszlo's description of the undivided wholeness of the Akashic Field. I have dropped the difficult cleansing portion of this session and begin the account after the breakthrough into its ecstatic portion.

S 21—Dying into Oneness

Then in the distance I began to see something. It kept expanding, from our solar system to the galaxy to the cosmos itself. It was the physical universe and the underlying forces that built and sustain the Universe. It was something both physical and archetypal. It was not a symbolic representation of the Universe but the real thing. It was continuous with the universe I had experienced on the Cosmic Tour, but many times larger and more complex. It was beautiful beyond words and absolutely captivating.

As I expanded into what I was seeing, I was becoming larger. I learned by becoming what I was knowing. I discovered the Universe not by knowing it from the outside but by tuning to that level of my being where I was that thing. All I can do at this point is to sketch the highlights of the experiences that followed, which will not do justice either to their cognitive structure or to their experiential intensity.

What stood out for me in the early stages was the interconnectedness of everything to form a seamless whole. The entire Universe was an undivided, totally unified, organic whole. I saw various breakthroughs—quantum theory, Bell's theorem, morphogenetic field theory, holographic theory, systems theory, the grand unified theory—as but the early phases of science's discovery of this innate wholeness. I knew that these discoveries would continue to mount until it would become impossible for us not to see the Universe for what it is: a single unified organism of extraordinary complexity and subtlety reflecting a vast Creative Intelligence—the Mind of the Universe. The intelligence and love that was responsible for what I was seeing kept overwhelming me and filling me with reverential awe.

The Unified Field that was underlying my physical existence completely dissolved all boundaries. As I moved deeper into it, all borders fell away; all appearances of division were ultimately illusory. No boundaries between incarnations, between human beings, between species, even between matter and spirit. The world of individuated existence was not collapsing into an amorphous mass, as it might sound, but rather was revealing itself to be an exquisitely diversified manifestation of a single entity.

As my experience of this seamless Universe progressed, I came to discover that I was not exploring a universe "out there," as I had in session 19, but a universe that "I" in some essential way already was. These experiences were leading me step by step into a deeper embrace of my own reality. I was exploring the universe as a dimension of my own existence, slowly remembering aspects of my being that I had lost contact with. This exploration seemed to answer a cosmic need not only to know but to be known.

Initially I was on a cosmic tour not unlike session 19 when I realized again that this larger field of consciousness that I was with (or in) had been waiting a long time to be recognized. Again I began to weep as I felt its heartfelt longing to be known. Then I asked something I had not asked before. I asked, "Who am I talking to?" With that question my experiential field began to change, and I dropped into a new level of reality. It was as though I fell into a deeper operational level where I discovered that I was, in fact, with MYSELF. The creative impulse that had been "other" to me at the previous level was at this level myself.

This mysterious progression repeated itself many times and in many variations. It continued for hours. I would be at one level of reality far beyond physical diversity, and as I sought to know this reality more deeply, I would experience a kind of dying, a falling away, and would slip into a new level where I would discover that this duality, too, was but another facet of Myself. Over and over again in detailed progressions, I was led to the same fundamental encounter.

No matter how many times I died or how many different forms I was when I died, I kept being caught by this massive SOMETHING, this IT. I could not leave IT, could not escape IT, could not, not be IT. No matter how many adventures I had been on, I had never stepped outside IT, never stopped being IT. There simply was no outside to My Being. There was no other in existence.

As I moved into these levels of increasing ontological simplicity, I entered a profound stillness that reawakened a distant, vague memory. "Where have I known this before?" By following this stillness, I was guided back to what seemed like a time before creation, back to the ontological fount of creation. In this stillness I was "with Myself" in ways that I had been long ago, but not for billions of years. It was a time of reunion, a time of being whole after a terribly long separation. . . .

I continued to ask my questions:

"What is happening here?"

"How does this work?"

"What has it been like for you?"

And with each question my experiential field changed, opening me to one cosmic process after another. I cannot describe these experiences adequately because the categories of thought derived from space-time do not lend themselves to remembering clearly or translating into words experiences of realities that lie outside space-time. Though my ordinary waking consciousness is being gradually changed by these experiences, it is still too cognitively restricted to be able to hold on to them in sufficient detail. What I experienced, however, repeatedly swept me into ecstasy.

"Amazing!"

"So that's how that works!"

"Oh, goodness!"

"How much do you want to see?" I was asked.

"More!" I answered, and always more would unfold. It kept unfolding for hours.

Experiment #4. *Session 36*

The next experience comes from several years later and a different phase of my work. In this phase, my consciousness was stabilizing at the causal level of consciousness, and I was being taken into different modalities of Oneness, as though to help me internalize this core truth more deeply. In the thirty-sixth session, I entered a vivid experience of what Buddhism calls emptiness (*śūnyatā*). Buddhism teaches that all conditioned reality is characterized by emptiness of self. When one experiences emptiness, one experiences the complete absence of an autonomous, independent self, not only in one's own person but in all reality. As I experienced it, Oneness and emptiness are two sides of the same coin. When we experience the innate Oneness of the living universe, it becomes clear that separate, independent "things" do not exist. Conversely, when the sense of being a separate self dissolves completely, the Oneness of life spontaneously rises in our awareness. Laszlo describes Akasha as the seamless field that underlies and permeates all differentiated existence. In this session, it was as though I was experiencing the forms of physical life from the perspective of Akasha.

S 36—the Forest

This session represented a radical jump in experiential categories that was absolutely intoxicating. New experiences and a new way of thinking opened today.

The theme of the opening movement was "If only you could see reality as it truly is!" Layer after layer was being pulled away as I moved beyond physical existence and the categories of physical existence. As in the past two sessions, there was no pain and no chaos, only a pervasive sense of disorientation. It felt as though I had actually taken a smaller dose of LSD. How could I be experiencing so many dramatic transitions and yet still be so centered? Assuming that the energy of today's session was as intense as in previous sessions, where was this energy focused?

In pursuit of answers to these questions, I turned to address that small, stubborn remnant of unlovability I had been processing in the previous two sessions. I don't know how or why, but soon after focusing on this problem, it suddenly opened and everything changed. There was no explanation; no content emerged for me to examine. I simply moved quickly, quietly, effortlessly, into a new reality. I entered a world that moved according to completely different patterns than any I have known before.

In this entire world, there were no separate "units," no "things." There was diversity and movement, color, shape, and form, but no isolation. Life here was not the assemblage of parts but the harmonious expression of a unified whole rippling through life, the way wind ripples through a wheat field. It was as though the wheat knew there were no separate stalks but only one "wheat" coming into expression through millions of strands.

The vision that mediated this experience was that of a Forest— thousands of trees and grasses, hundreds of species rendered transparent to the universal energies of soil and sun moving through them. Diversity did not rupture oneness. Oneness expressed itself in diversity without itself falling into diversity. "If only you could see reality as it truly is." As I saw it today, reality was a fluid energy expressing itself in diversity. The trees were many, but a single life force flowed through them all, ignoring completely whether they were pines, oaks, or maples. This underlying energy brings everything into existence, keeps everything alive, and reabsorbs everything at its end. It knows no divisions whatsoever. My usual divided reality was completely unreal. Division truly is an illusion.

The reality I was in preserved the forms of Earth. There was activity, process, movement, and people. Everything was as it normally is except that no "I" was present anywhere, either in me or in any other. I kept thinking to myself, "Nothing has changed except that I'm not here. 'I' do not exist." What a cosmic joke! What a relief! What freedom! Here was my form, my processes, my distinctive patterns of awareness, but there was no "I" to be found. It was simply gone. Feeling but no feeler; thinking but no thinker. What quiet, expansive joy. It continued for hours.

Experiment #5. *Session 45*

The personal and cosmological vistas in the four sessions presented up to this point have had content that allows them to be described in language we can recognize. They have shapes, actions, and forms. As one moves into deeper levels of awareness, all these drop away, making the experiences harder to describe. The joy they release and the learning they bring come not from the things seen but from the quality of consciousness itself. There is a sense of entering purer and purer states of contentless awareness. Often, the vehicle of these encounters is Light.

Many consciousness researchers have reported being flooded with a supernatural radiance after going through ego death. As one moves into still deeper levels of consciousness, the quality of this Light changes. It becomes clearer, more intense, more luminous. Each step beyond matter, beyond the soul, beyond the collective psyche, and beyond even archetypal reality takes one deeper into a living ecology of Light. Eventually, one discovers that the entire universe floats in an Ocean of Radiance. If the Akashic Field takes a form at all, it may be this transcendental Light. Science may describe the physics of this Light; what the following session contributes is the experience of be-coming Light.

In the last five years of my journey, I began to enter an exceptionally clear, exquisitely ecstatic Light that I called the Diamond Luminosity. Buddhism calls it *dharmakaya*, the "pure light of absolute reality." Its clarity was so over-whelming, its energy so pure, that returning to it became the sole focus of my remaining sessions. I entered this Di-amond Luminosity only four times in twenty-six sessions, signally the rare nature of these encounters. The session that follows was the first of these four encounters.

S 45—the Death State

Today is a day I have waited many years for. How can I express my gratitude to every person and circumstance that made today possible? How can I say thank-you enough?

After a long period of opening, I found myself repeatedly saying, "I have earned the right to die." Far from fearing death, I was seeking it out, demanding that death come to me. I was deflecting half measures and insisting on my right to a complete and final death. I had done my work; I had earned the right to die, and I was calling on this right. My litany focused me, carrying me deeper and deeper to a point of complete concentration.

From this position of absolute focus, I began to die. Oh, what sweet death! I began to savor what was happening. What I had previously feared now opened as incredible sweetness. How wonderful to experi-ence death! What a surprising reversal! Thank you, thank you.

Upon dying, I moved into an ecstatic mode of experience that was different from anything I have previously known; the entire flow of the experience was different. Light-filled, yes; a universe composed of noth-

ing but light. But what stood out for me is something I have difficulty finding words for. It was as if I had moved inside the inner flow of God's being, as if my life was now bending and flowing through a being of infinite dimensions. There was nothing amorphous or fuzzy about the experience; on the contrary, it was extraordinarily clear and precise. The boundaries of this clarity exceeded anything I had previously known.

Apparently, one death was not enough to get the job done in my case. I found myself standing in the middle of a circle of spinning bands of colorful energy that held my entire life. All my time moments were present in them. I fell into this circle, touching some part of my life, but as soon as I did, it "died out from under me," and I instantly found myself in the luminous death state beyond individual identity. Then I would be returned to the center of the circle and the process would repeat itself, my now falling in a different direction and touching a different area of my life. Over and over again, I went through this process of "dying in all directions," driving home the point that there was nothing unfinished here. Wherever I turned, there was no resistance, only effortless death and incredible sweetness.

The repetition kept expanding the scope of the transition, taking me deeper and deeper into ecstasy until eventually there was no center to return to, only the pure, seamless condition of the death state. What strange language to describe our true nature!

The Death State.
Incredibly clear.
Luminous beyond measure.
Incredible age.
A seamless intelligence running not above but
    inside existence.

Reaching out and moving into larger wholes of
experience.
Ecologies of experience encompassing thousands,
perhaps millions of beings.
Human experience folded into Earth experience.
Just touches, tastes.
Ecstatic reverence for the integrated movement of
life throughout the universe.

For hours, I was carried along the currents of this condition. About this state, one says either too little or too much. The price of saying nothing is to risk forgetting the subtler textures of the experience, yet to speak creates the illusion that words are adequate, and they are not. Even after fifteen years, today was so unlike any previous mode of experience that language truly fails. Silent appreciation seems the best recourse, combined with ceaseless prayers of thanksgiving.

How can something so crystal clear,
so devoid of earthly form,
evoke tears of homecoming?
What are we that such imprisoned splendor,
once released, floods us with rivers of gratitude?
Whom shall we thank for what we are?
Where do I direct my deep appreciation?
There is no one place,
so I send my prayer into the seamless fabric of
existence,
left and right, high and low,
in infinite dimensions—all around.

My attempts to describe the experiences keep breaking down, and I end up repeating the same words over and over.

I am home. And free. And Light.

There is nothing more I can say.

THE VIEW FROM SCIENCE (II)

What Are Akashic Experiences?

We now return to the view from science. Our question regards the nature of our experience of the Akashic Field. This field, we have said, is real—as real as any aspect or element of physical reality. It is how the universe appears to us, its embedded observers. It appears as a universal information-and-memory field, governing life and healing in us and around us, and orienting evolution in the world. The question is, can we *experience* this field?

Let us first consider the typical features that mark an Akashic experience. This experience is a lived experience that conveys a thought, an image, or an intuition that could not have been transmitted by the five bodily senses, either because it is beyond their scope by reason of time and distance, or because it is intrinsically unexperienceable by a human intellect. Does this experience tell us something important about the real world?

Countless people of the widest variety of interests and

backgrounds have recounted the Akashic experiences they have had. These experiences seem to come to many and perhaps to most people. They come in many sizes, forms, and flavors. They can occur in prayer or religious exaltation, in information-based healing, or when the organism is near the portals of death. Typically, they are so realistic that the people who undergo them do not doubt that they're veridical experiences of the world.

What brain science can tell us

Akashic experiences are highly diverse, but they share a common feature. They suggest that the experiencing subject is not separate from the objects of his or her experience: "I, who has this experience, am linked in subtle but real ways to all people around me, and to nature, and to the universe as a whole." Typically, this is not a complete surprise; it is an "aha experience." *So, then, what I have experienced (or dreamed or visioned) is not just imagination after all. It has something to do with reality.* What has science to say about the truth of this intuitive realization?

What science can tell us is indirect, a conclusion to be gathered on the basis of various strands of evidence. Such a conclusion could be important in today's world because personal insight is not a sufficient reason for most people to accept "Akashic" experiences as experiences of the real world. They would ask for the testimony of science. Science needs to substantiate that Akashic experiences are experiences of the real world. It can do this by showing that the human brain is capable of processing and conveying information that does not come through the bodily senses.

Work at the frontiers of quantum brain research, supported by quantum physics and quantum biology, suggests that this is not impossible. There could be experiences that are processed by the human brain even if they are extra- or nonsensory in part or in whole.

The human brain, with its stupendously complex and coordinated system of neurons, is more than we thought it was: it is more than a classical biochemical system. It is a "macroscopic quantum system"—an entity that acts as a quantum system although it is of macroscopic (that is, of larger than quantum) dimensions. If this is so, information reaches the brain that originated beyond the bounds of the eye and the ear, and other sensory organs. The currently popular mainstream concept recognizes this, but maintains that such information cannot be processed by our brain in any way that they could reach our consciousness. The brain, it is said, is a biochemical system that receives, decodes, and transmits information from the senses as impulses conveyed by the nervous system. The information that is conveyed to the brain comes from the external world; more exactly, from sensory receptors: the eyes, ears, nose, palate, and skin. The mainstream conviction is that every thought, intuition, image, or experience that is not conveyed by the five senses is fantasy. It is merely an imaginary recombination of data received through the senses.

If the brain is truly a macroscopic quantum system, this limiting condition does not apply. As a quantum system, the brain receives information from the supersmall world of the quantum. How it does so is becoming better understood. There are structures in the brain that are of nearly quantum dimensions, and these structures receive and send information based on quantum connections. The quasi-instant, space-and-time-transcending processing of information is a basic feature of the brain and nervous system of the living organism. The information received and processed by the living brain is not limited to the here and now. It is "nonlocal."

Nonlocality is another finding that contradicts the currently dominant belief. The current concept is that information is associated with specific points in space and time and is limited to such points. This is "local realism." Local realism is based on the assumption that physical effects are local, and if they propagate, they do so at a finite velocity in space and time. They have definite, insuperable spatial and temporal limits. The propagations diminish and ultimately vanish with increases in distances and in time. If this is the case, the human brain receives information limited to specific points in space and in time. But there is evidence that this is not so. The brain also receives, and works with, nonlocal information. It processes such information. It is not a conventional biochemical system, but a macroscopic quantum system.

The information received and processed by the brain is not local, but nonlocal. All information propagated in space and time is nonlocal. This is experimentally confirmed. As already discussed, repeated experiments show that particles that at any time occupied the same quantum state remain correlated over all finite distances. Changes in the state of one part of a split particle instantly result in changes in the other part, even when they are distant and not connected in any conventional way. Spatial separation, and separation in time, are irrelevant to this correlation: the split parts can be anywhere and could have come into existence at any time.

Space-and-time-transcending correlations (Erwin Schrödinger called them entanglements) occur when quanta are in coherent states. In their pristine state, prior to an interaction, quanta are in such coherent states; special conditions have to be produced to break apart their coherence. However, when quanta are subjected to interaction (and already their observation constitutes an interaction), they become "decoherent." They assume the characteristics of ordinary macroscale objects.

It appears that interactions in the living body, even at the relatively high temperature of warm-blooded animals, do not destroy the coherence of the organism. This experimental finding was a surprise. Previously physicists thought that the random, heat-induced "Brownian motion" of particles in the body makes them decoherent: they become classical objects not manifesting entanglement. But research (by physicists Kitaev and Pitkänen, among others) shows that the problem of "heat decoherence" is not insuperable. Specifically organized networks of quantum particles—for example, networks where the particles are "woven" or "braided"—appear to be sufficiently robust to maintain quantum coherence even at ordinary body temperatures. As evolutionary theorist and sociologist Talcott Parsons said, "Braiding is robust: just as a passing gust of wind may ruffle your shoelaces but won't untie them, so data stored on a quantum braid can survive all kinds of disturbance."

The physiological structures that receive and process information in the brain are part of the cytoskeleton: they are of quantum dimensions. Proteins in the cytoskeleton are organized into a network of microtubules connected to each other structurally by protein links and functionally by gap junctions. The microtubular, subcellular network contains vastly more elements than the neuronal network. The human brain consists of approximately 10^{11} neurons and 10^{18} microtubules, which means that the microtubular network has not ten or a hundred, but ten million times more elements than the network of neurons. With filaments that are just five to six nanometers in diameter—the so-called microtrabecular lattice—the microtubular network functions at or close to quantum dimensions. Neuroscientist Ede Frecska and anthropologist Luis Eduardo Luna noted that the cytoskeleton's ultramicroscopic networks are likely to be the structures that perform the computations that transform quantum-level signals into brain-decodable information.[27]

It appears that two forms of reception of information from the world are available to the brain, and not just one. In addition to perceiving the world through the bodily sense organs, the brain can perceive some elements of the world nonlocally, through quantum decoding. This mode of perception, called by Frecska and Luna the "direct-intuitive-nonlocal mode," is just as real as the "perceptual-cognitive-symbolic mode"—ordinary sensory experience.

An analogous conclusion has been reached by transpersonal psychiatrist Stanislav Grof. He wrote, "My observations indicate that we can obtain information about the universe in two radically different ways. Besides the conventional possibility of learning through sensory perception, and analysis and synthesis of the data, we can also find out about various aspects of the world by direct identification with them in altered states of consciousness." Grof concluded that each human being is a microcosm that accesses information about the macrocosm, the universe.

The query regarding the realism of extra- or nonsensory—"Akashic"—experiences has a positive answer. The brain and nervous system can pick up information from the Akashic Field beyond the frequency range accessed by the senses. Beyond that range, for the brain to process information in a way that it can reach consciousness, calls for entering an appropriately modified "altered" state. Information could reach consciousness also in the ordinary aware state, but this is relatively rare: it is a vision, a hunch, or possibly even an explicit message that enters the flow of conscious awareness. Akashic experiences ordinarily happen in altered states. They can be catalyzed by meditation, exaltation, the contemplation of art or of nature, approaching the portals of death—and even, as Christopher Bache reports in the foregoing chapter, by psychedelic substances.

PART 3

GUIDANCE BY THE LAWS OF THE AKASHIC FIELD

THE WAY TO A BETTER LIFE
AND A BETTER WORLD

Connecting contemporary science with classical wisdom is not merely an intellectual exercise. True, it is aimed at increasing our understanding of the world and of our role in the world, but it is also aimed at eliciting reliable guidance toward a better life and a better world.

To find our way toward a better life and a better world we need science as well as spirituality. Science without spirituality misses the intuitive elements of human experience, elements that many great scientists have valued on a par with, and even above, reason and logic. But spirituality without science cannot offer reliable guidance for confronting the problems we face in the world. We need both science and spirituality, and we need them together, coherently linked. The need is for a dedicated and lasting alliance between science and spirituality. The feasibility and potential fruits of this alliance are the topic of this concluding part.

THE NEED FOR GUIDANCE

We are in urgent need of guidance. It is common knowledge that we are now an endangered species. The coronavirus pandemic of 2020/21 makes this clear. And yet this virus is not the only global problem we face. A number of local and global challenges await our urgent response. Here is a brief and by no means complete litany of the nature of the most urgent among them:

- We are stripping life from our oceans and replacing it with hundreds of millions of tons of plastic waste.
- We are cutting down our forests, exhausting our freshwater aquifers, and losing our vital topsoils.
- Our pollution is driving climate change that causes heat waves, droughts, and wildfires that shred the fabric of life on the five continents.
- Toxic industrial chemicals are flooding the environment of the big cities.
- Millions of newborns die of malnutrition in the first few months of their life.
- A billion or more human beings struggle just to stay alive, fighting thirst and imminent starvation.

The scope and the depth of such threats to life on the earth are getting wider and deeper with each passing day. Big business has a major share in producing them. The following are among the "collateral damages" produced by the profit-driven aspiration of business leaders:

- Workplaces that create poor health and generally miserable conditions for employees.
- Continued child-labor exploitation as well as gender and sexual discrimination.
- Long-term exposure to toxic chemicals, air pollution, and water contamination.
- In some poor countries, indenture contracts that keep families poor for generations.

The scale and magnitude of these problems have reached global proportions. In the last one hundred years alone, big business has altered the earth's climate, contributed to a massive decline in biodiversity, converted rain forests to croplands, added fixed nitrogen to soil and water, and disrupted the phosphorus cycle at levels that are catastrophic for human health.

We are in the final stages of the Industrial Age. This was an amazing age; it brought not only problems and degeneration but veritable miracles on the scene. We have gone from dirty coal to clean and cheap solar power; from traveling from New York to London under sail in five weeks to the same trip by air in five hours; and from local snail mail to global exchanges and telecommunication. We have gone from an annual world output valued at less than US $200 *billion* to an economy that exceeds US $80 *trillion*. Wealth has been constantly accumulating, but it has done so mainly in the coffers of the rich. The gap between the rich and the poor (the so-called Gini coefficient) is growing in all parts of the world, both within and between countries.

Human numbers have grown from less than one billion in the nineteenth century to 7.7 billion today. We know that each human requires food, water, and shelter just to survive, yet we are adding seventy-five million humans to the world per year. The resources required to sustain human life have been expanded by science and technology, but they are not infinite. Ultimately the point will be reached where the limits of resource expansion are attained. We are already nearing that point. In regard to some essential resources, we are scraping the bottom of the barrel.

We need to find a better way to go—a way that can ensure the sustainability of life for all humans on the planet.

THE ROLE OF VALUES

We *need* to change, and we *can* change. Whether we will change in time depends on us. Do we know what is involved in changing? Do we know *how* to change?

The crucial question is to know what makes human life healthy and sane, sustainable and flourishing. What makes it different from a way of life that is "dis-eased" and "insane"? The answer is, our *values*. We are not automatons that blindly obey preestablished instructions. We have a degree of freedom, and that freedom includes having and choosing our values. Values project ideals, and when we live by our values, we attempt to attain the ideals they project. Genuinely held values are powerful factors motivating thinking and conscious acting in the world.

The question to consider is, What are the values that

project healthy and sane ideals, capable of leading us toward a better life and a better world? Pioneering values researcher Karin Miller came up with a list of ten such values. We list them below and summarize the ideals they project.[28]

1. *Unity.* Together, we make up one integral system of life on the planet. Our diversity is not an obstacle to unity, but the means for achieving it. Viable systems combine diversity with unity: they find their unity through the coordination and cooperation of their diverse elements.

2. *Community.* We are whole-systems, and when we act in isolation, we are not effective. We can only unfold our potentials when we act as a community, forming a greater unity through the coordination of our diversity.

3. *Life.* The emergence of life on the earth was momentous. The evolution of the biosphere from a sea of biochemical elements was a stupendous occurrence, similar in significance to the Big Bang. The safeguarding of the life that emerged on this planet needs to be the ultimate aim of all the things that emerged and continue to evolve in our biosphere.

4. *Freedom.* Freedom means having real alternatives before us to pursue the path we select for assuring and evolving our existence. This freedom is present throughout the realms of life, and it grows with the level of evolution reached by living systems. In humans, freedom reached

the point where we have a choice between survival and extinction—between living in the biosphere, or opting out and going under.

5. *Connection*. Because all things in our world have an
impact on all other things—economies, cultures, environments, political systems, and our own life—
interconnection is basic to human life, and to all life.
Maintaining it is an instrumental value, leading to
the realization of other values, but it is critical and
fundamental in its own right.

6. *Sustainability*. Human life on both the individual
and the collective level can be rendered com-
patible with the rhythms and balances of other
forms of life and the physics and the biology of
the biosphere. Doing so ensures and even pro-
longs human life in harmony with all life in the
biosphere.

7. *Creativity*. For a conscious human being to exist is to
create—to express himself or herself through whatever medium he or she wishes. Allowing people's
creativity to unfold is as basic to their existence as
the air they breathe and the water they drink.

8. *Empowerment*. We need to be empowered to live and
thrive on this planet, and because we are interconnected, empowering ourselves empowers the whole
human family.

9. *Choice*. Choice is the precondition of the effective
exercise of freedom and is basic to human life.
Maintaining, and within the limits of possibility en-

larging, the range of our choices is a healthy and sane aspiration.

10. *Integrity.* In the biosphere, all things are integral systems, or parts of such systems, humans included. Maintaining our physical and biological integrity and matching it with the integrity of our mind and consciousness is a necessary condition of health and sanity for us and for others.

The above values project constructive, life-sustaining ideals. They are values we can embrace. But we have the freedom also to neglect and dismiss them. Values are social and cultural constructs and not eternal truths—they can be ignored, changed, and replaced. Those we embrace affect us and our societies. They define the character of our civilization.

It is in the most immediate and fundamental human interest to change the values of a critical mass in contemporary society. Embracing the now-dominant values has produced highly undesirable conditions. We need to embrace values that make for health and sanity in our life and in our world. They include the ten values cited above: the values of unity, community, life, freedom, connection, sustainability, creativity, empowerment, choice, and integrity. They also include other values—each of us can add those we hold paramount. When we are guided by positive values, we are in tune with the laws that hold sway in the universe—we are coherent with all the things and beings that inhabit the earth.

TUNING IN TO THE AKASHIC GPS

How do values relate to laws—including the laws of the Akashic field? Values and laws are distinct entities. Values are social and cultural constructs, while laws are immutable givens in nature. But values and laws operate in the same world and are linked in practice. The values we adopt are not determined by the laws that ground our existence, but they are also not detached from them. The health and sanity of people, and of entire communities and societies, are strongly dependent on their coherence with the laws that govern existence on the planet. The laws of the Akashic Field are clear examples. They interconnect all things, conserve all things, and inform all things with a subtle but real drive toward coherence. Taking these laws into account is an essential factor in our choice of healthy and sane values.

Living in harmony with the Law of Coherence is especially important. Measuring our steps against the cosmic impetus toward coherence offers a reliable guidance system, a kind of cosmic GPS. It tunes us to coherence within us, and between us and others around us—and with the whole rest of the world.

It is in our best and most enlightened interest to live in harmony with the laws that govern existence in the universe. We need to return to universal harmony; we have failed to abide by this law in many ways. Modern technological societies became more and more divorced from living nature. There is an urgent need for us to reconnect. Traditional civilizations such as the Hellenic, the Arab, the

Persian, the Indian, the Chinese, and the Japanese were far more in tune with this Akashic law. Not surprisingly, they could maintain themselves for thousands of years.

Even if the striving to live in accordance with Akashic laws has been abandoned in modern industrial societies, the impetus for it did not disappear. The universe's guidance system—a kind of cosmic GPS—continues to exert a subtle influence. This influence becomes tangible when we enter deeper into our own consciousness, in altered, meditative, or exalted states. Artists, poets, and children do not need to make special efforts to achieve the consciousness where they receive guidance from the cosmic GPS. When a normal child is told about the nature of a healthy life, he or she is likely to say, "Yes, I know that." Modern adults may need more convincing. If willing, they can take account of leading-edge science's corroboration of traditional Akashic tenets. And they can also enter the deeper states of consciousness in which they themselves receive tangible guidance from the cosmic GPS.

Consciously and purposively tuning us and our highest values to the laws of the Akashic Field makes good sense. If we live in a world where all we do influences and is influenced by what everyone else does, it is foolish to do anything that harms anybody. Whatever harms one, harms all— and harms whoever caused the harm. In such a world, it is essential to realize that whatever we have done in the past did not vanish, but continues to influence all we do in the present. And will influence all we will do in the future. It is also foolish to ignore the great trends that unfold around us and inspire the deeper domains of our consciousness. If we fail to live consistently with these trends, we reduce our chances of creating a better life for ourselves and a better future for all.

The greeting exchanged by young people makes excellent sense: "May the force be with you." The force is there in the universe, and it is in us and with us. We need only to be aware of it so we could head toward a better life and a better world.

ABUNDANCE:

An Epilogue

Jean Houston

To read this work, the potent explorations of Ervin Laszlo and the visionary essays of the contributing authors, is a journey of radical and radiant discovery. To use the ideas offered here is to adventure into living from the *Abbondanza* that is the very nature of reality and the source and resource of all that lies within us, even if unknown by us.

Abbondanza! was the favorite expression of my Sicilian grandmother, Vita Todaro. She would look over the groaning board of the feast of Sicilian dishes she had created for the Sunday dinner and cry out with joy and widespread hands, *"Abbondanza!"*—to which the twenty or more of our relatives would all cheer and respond, *"Brava, Nana!,"* as we happily dug into the six-hour-long celebration of feasting and family connecting. But then at night on the roof, I would see my grandmother, her head thrown back, calling out her appreciation to the star-splashed sky: *"Abbondanza!"*

The next day, taking me to the park, she would peer into a baby carriage and whisper to the little inhabitant, *"Abbondanza."* That word, that celebration of life in all its forms and wonders, has informed much of my life and what I believe and teach. And that is what I discover in this remarkable work of new science and its application to the meaning of our existence. It could also be another word for the Akashic Being, which supports, sustains, and generates the *Abbondanza*, the embodiment, the creativity, and, yes, the abundance of our reality. Rumi called it "the treasury of unseen generosity."

We live in a time in which the vibrant *Abbondanza* of our inner nature requires to be discovered, explored, and brought into our lives. And this is where the Akashic Field rises and enters into our lives. It tells us the most important thing we can ever know: we don't just live in the universe, the universe lives in us!

Among the more conscious members of the human race, this has long been suspected and put in parables and evocative sayings. My favorite saying is one from the Gospel of Thomas, where Jesus tells us that if you are looking for the kingdom of heaven in the sky, the birds will get there before you. If you look for it in the seas, the fish are there before you. Rather, the kingdom is spread out before you. If you read *Akashic Field* for "kingdom," you have a revolution in thinking and being, one that brings new mind to old matters and opens up the immense treasures of the universe to the hidden splendors that lie within each of us.

What is revealed in Ervin Laszlo's seminal thinking, and in the pages of this book, is an evolutionary event, a revelation that announces the "punctuated" equilibrium that attends evolutionary jumps. We are in Jump Time. Our job is to know this, and to participate in, make use of, its powers.

My own work in human and cultural development has given me a finger on the pulse of what is happening. Increasingly, having worked in over a hundred countries, I sense a new transcultural music coming into time. A welter of multiple meters and offbeat phrasing, it is a coding of creativity and imagination, a counterpoint of styles of knowing and being. As cultures come together and exchange their essence, in the light of the new sciences, they join in a cadence of awakening, a new idiom of consciousness as exhilarating and revolutionary as it is always awesome and sometimes fearful.

This rhythm of awakening carries us from the ballads of local concerns to the symphony of a larger ecology of Being. In this larger music, we feel the Pulse arising from the earth and the cosmos and come to understand that our individual life is part of the Great Life unfolding. The cosmos is finding welcoming recipients for the generosity of its gifts. Thus, in recent years I have extended my work in social artistry, to explore ways of bringing the implications of life in the quantum universe and in the Akashic Field to potentials that may up to now have seemed mythic rather than real.

What I offer here are tools for consciousness in the Akashic Field—how I have used these ideas to generate capacities in students and participants in my work. I do not pretend that these are scientific studies, but rather artful means to evoke latent capacities that both excite and invite profoundly new ways of being and doing, given our new understanding of our place in the universe.

Laszlo tells us that the Akashic Field is a cosmic information-and-memory field that interlinks and conserves all things throughout space and time. He adds that these claims are not merely metaphysical abstractions, but properties of the rediscovered fundamental nature of the world. They are supported by the findings of the contemporary sciences. In the language of science, Laszlo writes, they can be described as (i) instant universal interconnection among all things, (ii) universal and conceivably infinite memory, and (iii) interactive evolution tending toward a particular state, or set of states, in the universe. He assures us that this is not some fantastic sci-fi epic, but perhaps the clearest exposition of the nature of the universe, of the structure of reality itself. Reality is a domain where "everything connects and interacts with everything else." In every instant when this happens, our reconnection to reality is complete with infinite memory, creativity, and integral evolution toward oneness and coherence.

Since time immemorial visionaries have sought to understand that we live in the Larger Nondual Reality, in the Great Oneness, and in the many unseen worlds that coexist with our current existence. Whether as magus, mystic, shaman, or sage, the self within its own unbounded nature is basically identical with the quantum mind and therefore has many more capacities than those that operate in our local consciousness. Working with these concepts—both spiritual and scientific—have enabled people to be, to do, and to create in ways that mean a higher level of human achievement.

I often take my students into an exercise intended to shift them into the larger magnitude of their own identity. For example, I say, "Let us now explore the art of living in an expanded self, in an enchanted state which gives access to latent physical capacities and higher states of bodily functioning for health and well-being, and mind states that enlighten us as we reach the psychological states where we can consciously orchestrate our moods and emotions, our relationships, and the vast assembly of all the different and brilliant personalities and skills that we have within us. Be very present in your body as I speak to you now. Please know that your body is the stuff of stars and of the minerals of the earth. Your blood runs briny with the seas, the essence of oceans spills through your veins and arteries. And the sediments of earth make up your cells. Your genes are universes in themselves, coded with enough information to re-create the world. These elements of earth and sky, of nature and cosmos, compose your physical being and compose your consciousness: the inner mirror of the great reality that has pushed us to the choices we now face. It all has a purpose, even a destiny. You, my friend, are the apex of the 13.8-billion-year evolutionary process that has resulted in you and in your life."

Then, using Plato's understanding that everything and every person contains the eidos, the divine idea, I might evoke a sense of moving into our own unique eidos, our quantum blueprint (quantum healer Dr. Maria Sági calls it the individual's unique "morphic pattern" in the Akashic Field). Your responsibility—I can now add—is to reconnect with this inner nature, this emergent evolutionary nature, this optimal Akashic template. This quantum blueprint that has been waiting for you to recognize it and to be filled by it. "Come closer," you say, and it does. For each of you, its appearance is unique. You may feel it as light, as joy, as something that is filled with the codes of your higher nature. It has been waiting for you to sense and recognize as your higher destiny all these years, in your own local life, but also for millennia before that, ever since the first cosmic seeds began to bloom. And now, the moment has come. Move toward it as the Akashic pattern, the quantum blueprint, moves toward you. Feel yourself connecting with its energy, its wondrous plans for you. Connect with the energy and the plan of your larger life.

As the quantum template moves in you, know that you are becoming a superb catalyst, a carrier of new genesis vital for all of us as our world is getting ready to change and transform. Your mind is growing so you can think in many ways—in words, in images, in thought ways that border on genius. In fact, you yourself can become genius.

Much that may have been lost in you from your childhood and adolescence, together with those remarkable skills and qualities that were previously merely latent in you, are now becoming real . . . courage, passion for the possible, rigor, diligence, a wave tide of joy and belief, the creative life, your spirit inspired by God, the Cosmic Consciousness, the Supreme Beloved. You are loved, nurtured, empowered, called forth to your highest destiny at this time.

Given this entrance into our quantum blueprint, we can have the sensibility and openness to explore life in the Akashic quantum universe. A good place to begin is the new experience of time.

You can use altered states of consciousness to demonstrate ways of living in more fluid categories of space and time, categories that allow us to experience subjective time in a short amount of clock time, even though it is felt to be much longer. We are able to experience adventures, write books, finish projects, go voyage in the seas of the unconscious, even learn or rehearse things that would normally take a much longer time. "Close your eyes and breathe deeply, inhale, exhale, inhale, exhale, follow your breathing all the way in and all the way out . . . become more and more relaxed. Now I'm going to give you considerably more time than you need to do the following. You will have one minute of objective or clock-measured time, but with special time alteration, that one minute will be just as long as you need to live out a very interesting adventure. You may, for example, take a trip around the world, or go and visit a place you have long wanted to experience, or go and see old friends. But traveling around the world is always interesting. Now these experiences may seem to take a minute, a day, a week, a month, or even years, but you will have all the time you need in subjective time, because only a minute of clock time will have passed. I will hold the time. Begin your experience now."

After one minute, I call time and ask, "What did you experience?" Most people report that their internal experiences seem a great deal longer than one minute, and for some, time itself becomes meaningless. The most common reported experience is for people to take a trip around the world. I have been astonished at the lavish and lengthy descriptions offered by these one-minute world travelers. Once you are introduced to the world of alternate time and time expansion, such feats, evidently natural to subjective time, become part of normal experience. You learn to apply the technique to rehearsing skills, improving relationships, writing books, music, exploring challenges—the list is endless. (And, speaking personally, I can add that this is the only way I can get through my enormous workload.)

Similarly, from the Akashic perspective of the simultaneity of past, present, and future, we are able to enter into a minor unpleasant memory from the past and work imaginatively to change the story of that event until it becomes a realistic part of our living memory, with the old memory becoming a faded dream. (In some cases, people who were present at the original harsh happening begin to remember the subsequently re-created incident as that which had actually happened—such is the way of the interconnection of space and time in the Akashic Field.)

The potential of the Akashic Field allows us to experience the law of universal interconnectivity. We are familiar with the research that demonstrates how bonded particles remain connected; their quantum states are correlated. Ervin Laszlo reminds us, "Such correlations are not limited to the micro-domain of the quantum. Nonlocal connections have been discovered within living organisms, as well as between living organisms and the world around them. The universe as a whole manifests fine-tuned spatial and temporal connections that create a level of correlation in the state of the entities that appear in it."

In Laszlo's earlier book *What Is Consciousness?* I offered an example to demonstrate my conviction that "we are amphibious beings with regard to time and space, living in a cosmic hologram several lives of self and psyche at the same time." And if you add to this the conundrum of certain aspects of quantum physics that suggest that all possible futures are here right now, we realize that we stand at a crossroads in which we can select one path from a complex reality. This means that in our so-called unconscious lies not only the repressed or forgotten experiences of our life, but, dig a little deeper, you also have the experiences of ourselves in different dimensions of time and space. Jennifer in one reality has Lyme disease. Jennifer in another does not, so she visits this alternative reality or time and consults with the healthy version of herself. When she returns, she is inspired or impacted with the experience and immediately feels healthier and without the usual symptoms and fatigue. She then goes about seeking the kind of medical help that helps her to be rid of whatever aspects have remained of the disease.

Since my initial experiences, I have performed variations on the exercise of alternative selves with my students and have found similar positive results. Often the results have not had to do with the healing of illness, but rather with an increase in abilities in performance in arts and sports, enhanced creativity, positive psychological shifts, entrepreneurship, and leadership. This has made for a more fruitful exchange and sense of well-being.

We are players in a great game called Paradox. And what is the paradox? It is that we can be a child in Brooklyn sitting by a window knowing everything, and we can be the universe knowing the child. It is that we are both infinite and finite beings: As finite beings we are Godstuff incorporated in space and time. As infinite beings, we are the Living Universe in an eternal yet spirited form of Itself. As this infinite self-expressed aspect of God as a form of the Living Universe, we find ourselves capable of creating and sustaining an individual finite self that is us, human beings, who are the microcosm or fractal of the infinite self. Which means *that you don't just live in the universe. The universe lives in you.* The human selfing game may be what Infinity does for fun. Not realizing this, we live in galloping ambiguity, caught in a limited biodegradable space-time frame, yearning for our greater self.

What we do, where we go, whom we meet, is I believe to become who and what we were patterned to be. As we cannot shrink the infinite to fit into the finite, because if we do so, we just end up with a fundamentalist God, we can extend through conscious work on ourselves the capacity to expand and thus to enter into partnership with the infinite. Then, and this may be the goal of the Paradox game, we do indeed discover that we are an infinite self-creating and self-sustaining individual human self.

This is the Paradox of Partnership resolved. The game is to overcome the illusion of separation. The ideas contained in this book provide some of the finest and most coherent understandings of how we are all sustained and contained within the vast domains of the Akashic Field, in the field that we live and breathe in and in which we have our being. Here, as we connect the finest fruits of contemporary science with the most treasured instances of traditional wisdom, genuine spirituality and science meet. They marry and engender the reality and the practices that can evolve our self into the *possible human,* and with it into the *possible society* that is patterned in the field, and encoded in us and in all beings who share the field.

All the things and events that ever existed or happened continue to exist in the Akashic Field on the level of pure potentiality. As we think them, they arise from their Akashic base with other items of potential information. Together they become for us a field for new ideas, creative insights, and, ultimately, new ways of being. In this way the great field of Being lives and dwells and grows in our world. And so, the evolution of our life and of all life continues in many places and takes many forms. This has both direction and plan; it is in no way a random series of happenings.

I recall physicist Fred Hoyle's marvelous observation that thinking that the living organism came together as a result of random, unplanned happenings is like saying that a hurricane could sweep through a junkyard and create a working airplane! There is a Great Mind, an *élan vital,* that orchestrates it all to higher levels of complexity and consciousness. Thus we are the receivers and the partners in the great game of evolution whose motto is *Abbondanza!*

Here and now, in a new millennium, we stand on the shore of time, at the edge of history, receiving the winds of change. We welcome the homecoming of the spiritual force that can quicken our passion for life in this Jump Time. I have known for a long time that in this holographic universe, where the quantum reality is emerging into humanity's consciousness, what we envision and become in person impacts on the human collective in profound ways. Because we are one with the collective as well as being a personal self, I have worked internally to change myself at the same time as I work to write the various musicals that could support a transformation on the physical level. My life's mandate has been a slight variation on a quote by James Joyce: "I go to forge the uncharted course of my race in the smithy of my own soul."

Fears abound. Few, if any, have been trained for such a time. We feel cluttered and burdened with the learnings of a world now passing into history, shackled by beliefs birthed out of a narrower view of the cosmos. No matter how "postmodern" we pretend to be, each of us has been marinated in the medieval soup of the mind. To face the radical needs of the future, we need a new natural philosophy, one that encompasses an appreciation of our own evolutionary possibilities. I believe that this new philosophy has been birthed in the pages of this book, inspired by the luminous and prodigious mind of Ervin Laszlo. He dared, in the words of T. S. Eliot, to "redeem the time. Redeem the unread vision of the higher dream."

ANNEX:

A Science-Based Document Connected to Spiritual Wisdom

In the year 1996, the present writer served as chairman of the Advisory Board of the Auroville International Foundation, headquartered in Auroville, near Pondicherry, in India. In the late spring of that year, on the occasion of his work with the board in Auroville, he met with the Dalai Lama, who was on a visit to nearby Chennai (called Madras at the time). Given the heavy commitments on the Dalai Lama's time, we had a meeting scheduled for just ten minutes. But that meeting was spontaneously prolonged for the rest of that day; His Holiness canceled all his other appointments. Accompanied by his first secretary, who took meticulous notes, he reviewed the document the writer had brought along. He began to read it, and a meeting that was scheduled as a brief encounter became a several-hours-long intense workshop.

The document the writer presented to the Dalai Lama

was called a "Manifesto on Consciousness." The writer drafted it on the basis of the information provided by various scientific sources. It was to be a call on behalf of the science community to the general public, to wake up and take notice of trends and developments that had to be changed if we were to avoid a major crisis. As His Holiness began to realize what the document intended, he began to give it deep thought. He kept adding his ideas to it, paragraph by paragraph. In this way, a document that was a product of science-based information became the spiritually sourced "Manifesto on the Spirit of Planetary Consciousness." The closing section of the Manifesto, its impassioned call for a new "planetary" consciousness, is reprinted here.

The Call for Planetary Consciousness

In most parts of the world, the real potential of human beings is sadly underdeveloped. The way children are raised depresses their faculties for learning and creativity; and the way young people experience the struggle for material survival results in frustration and resentment. In adults this leads to a variety of compensatory, addictive, and compulsive behaviors. The result is the persistence of social and political oppression, economic warfare, cultural intolerance, crime, and disregard for the environment. Eliminating social and economic ills and frustrations calls for considerable socioeconomic development, and that is not possible without better education, information, and communication. These, however, are blocked by the absence of socioeconomic development, so that a vicious cycle is produced: underdevelopment creates frustration, and frustration, giving rise to defective behaviors, blocks development.

This cycle must be broken at its point of greatest flexibility, and that is the development of the spirit and consciousness of human beings. Achieving this objective does not preempt the need for socioeconomic development with all its financial and technical resources, but calls for a parallel mission in the spiritual field. Unless people's spirit and consciousness evolves to the planetary dimension, the processes that stress the globalized society/nature system will intensify and create a shock wave that could jeopardize the entire transition toward a peaceful and cooperative global society. This would be a setback for humanity and a danger for everyone. Evolving human spirit and consciousness is the first vital cause shared by the whole of the human family.

In our world static stability is an illusion; the only permanence is in sustainable change and transformation. There is a constant need to guide the evolution of our societies so as to avoid breakdowns and progress toward a world where all people can live in peace, freedom, and dignity. Such guidance does not come from teachers and schools, not even from political and business leaders, though their commitment and role are important. Essentially and crucially, it comes from each person himself and herself. An individual endowed with planetary consciousness recognizes his or her role in the evolutionary process and acts responsibly in light of this perception. Each of us must start with himself or herself to evolve his or her consciousness to this planetary dimension; only then can we become responsible and effective agents of our society's change and transformation. Planetary consciousness is the knowing as well as the feeling of the vital interdependence and essential oneness of humankind, and the conscious adoption of the ethics and the ethos that this entails. Its evolution is the new imperative of human survival on this planet.

The "Manifesto on the Spirit of Planetary Consciousness" was signed by the Dalai Lama and adopted at an international meeting of the Club of Budapest at the Hungarian Academy of Sciences on October 24, 1996.

NOTES

1. Swami Vivekananda, *Raja-Yoga* (Calcutta: Advaita Ashrama, 1982).

2. Henry Stapp, *Mindful Universe: Quantum Mechanics and the Participating Observer* (New York: Springer, 2011); Pierre Lévy, *Quantum Revelation: A Radical Synthesis of Science and Spirituality* (New York: Select Books, 2018).

3. For the sake of simplicity and clarity, we shall continue to spell *information* in the usual form as one word, except where it is used specifically as the impetus that "forms" the universe.

4. For references and related views see the author's *The Intelligence of the Cosmos* (Rochester, Vt.: Inner Traditions, 2017).

5. C. N. Yang and R. Mills, "Conservation of Isotopic Spin and Isotopic Gauge Invariance," *Physical Review* 96 (1) (1954).

6. Martin Rees, cited in Bruce Rosenblum and Fred Kuttner, *Quantum Enigma: Physics Encounters Consciousness* (New York: Oxford University Press, 2006), 193.

7. Pim van Lommel, *Consciousness Beyond Life* (New York: Harper Collins, 2011).

8. In a more technical vein, we should note that the information stored extra-somatically is not actual information, but information in the potential form. The act of its recall "actualizes" the information, creating it as an effective presence in space and time.

9. See B. J. T. Dobbs, *The Janus Faces of Genius* (Cambridge, UK: Cambridge University Press, 1991).

10. There is an intriguing question regarding the origin of the laws that inform our universe. Two hypotheses can be envisaged. One is that the laws have been transferred from the implicate order to the explicate order at the birth of our universe. In that case we do not and cannot know their origin in the implicate order. The alternative hypothesis is that the laws are the product of information generated in a prior universe. Information carried over from that universe (or universes) would have in-formed our universe at the Big Bang. The former hypothesis rests on the assumption that the universe we inhabit is a singular event, extending from its fiery birth in the aftermath of the Big Bang to its ultimate end either in a superdense quantum state, or in an infinitely expanding "dead" universe. The latter hypothesis provides a further explanation, but it presumes the cyclic iteration of universes with the transfer of information from universe to universe. This iterative universe-evolving/devolving process could account for the presence of order and structure in our universe. It has been suggested by a number of cosmologists and has been described by this writer in an earlier book, *The Whispering Pond* (Rockport, Mass.: Element Books, 1994).

11. These values were first put forth in my book *The Phoenix Generation: A New Era of Connection, Compassion, and Consciousness* (London, UK: Watkins Publishing, 2014).

12. Robert David Steele, *The Open-Source Everything Manifesto: Transparency, Truth & Trust* (Berkeley, Calif.: Evolver Editions, 2012).

13. An invented amalgam of *technosphere* and *noosphere*.

14. World Economic Forum, *Deep Shift: Technology Tipping Points and Societal Impact*, Survey Report, Global Agenda Council on the Future of Software & Society, September 2015, http://www3.weforum.org/docs/WEF_GAC15_Technological_Tipping_Points_report_2015.pdf.

15. Sri Aurobindo, *The Human Cycle: The Psychology of Social Development* (Twin Lakes, Wis.: Lotus Light Publications, 1999/1950), 263.

16. Duane Elgin, "The Buddha Awakening, Integral Expanding, and a Second Axial Age for Humanity," *Journal of Integral Theory and Practice* 9 (1) (2014): 145–54.

17. Szilard Hamvas, Monika Havasi, Henrik Szőke, Gábor Petrovics, and Gabriella Hegyi, "Different Techniques of Acupuncture—Part of the Traditional Chinese Medicine and Evidence-Based Medicine," *Journal of Traditional Medicine and Clinical Naturopathy* 6 (1) (2017).

18. There are several ways to convey coherent radiation of wavelength 633 nm to the diseased organism. Some stimulate the body itself to produce coherent radiation at this wavelength, and others convey this radiation from the outside. An example of the latter is soft laser irradiation, known as laser acupuncture. Today we know that in the course of defensive immune-system reactions at the phagocytosis stage, the body emits low-intensity light in the 633 nm domain without amplification to the domain of visible light. When we insert the acupuncture needle at one of the acupuncture points, radiation of wavelength 633 nm is produced, and this stimulates the entire system of meridians.

19. If we draw a line on paper or on the skin, we notice that it is embedded in a typical field distribution of frequencies. The ends of the line carry opposing polarities, and as a result, stationary electric and magnetic waves arise in the medium, and this pattern of waves continues beyond the line itself. If we change the shape of the line, for example, creating the letter *L*, the fields become correspondingly dense or thin around the angle in the line and produce corresponding shapes in the frequency distribution.

20. During the therapy, the subject reinforces the connection between the selected symbol and the traumatic information by *looking at a paper with his or her name, the exact date of the traumatic event, and the chosen symbol on it*. For more information, see Dr. Maria Sági with István Sági, *Healing with Information: The New Homeopathy* (London: O-Books, 2018).

21. The complete story of this journey is told in my book *LSD and the Mind of the Universe* (Rochester, Vt.: Park Street Press, 2019), from which these accounts are taken.

22. It is not easy to lift psychedelic experiences out of their original context, as I am doing here. Different levels of consciousness operate by different rules reflecting different patterns within the larger whole and therefore require careful introduction, for which there is not space here. I will simply explain where these sessions occurred in my larger journey and offer *LSD and the Mind of the Universe* as the more complete account.

23. I place my session accounts in italics to clearly differentiate the original experiences from their subsequent discussion.

24. I eventually came to understand this field of collective suffering to be memories of unresolved historical trauma aggregated in the collective psyche.

25. I have often wished that I had advanced training in physics and astronomy, for then I might have been able to retain more of what I was shown in

this and other sessions. The content was not inherently ineffable, but it was extraordinarily sophisticated and technical.

26. I use Stanislav Grof's categories of psychic, subtle, and causal levels of consciousness to identify the specific "platforms of experience" that opened on my journey.

27. For the documentation of these findings see inter alia the theories of Stuart Hameroff, *Ultimate Computing* (Amsterdam: North-Holland, 1987); Alexei Kitaev, "Quantum Error Correction with Imperfect Gates," in *Proceedings of the Third International Conference on Quantum Communication and Measurement*, ed. Osamu Hirota, Alexander S. Holevo, and Carlton M. Caves (New York: Plenum Press, 1997); Paul Parsons, "Dancing the Quantum Dream," *New Scientist* 2431 (2004): 31–34; Roger Penrose, *Shadows of the Mind: A Search for the Missing Science of Consciousness* (Oxford, UK: Oxford University Press, 1996); and Matti Pitkänen, *Topological Geometrodynamics* (Frome, UK: Luniver Press, 2006).

28. Karin Miller, *Global Values: A New Paradigm for a New World* (New Castle, Del.: Our New Evolution, 2015).

About the Author

Bernard F. Stehle

ERVIN LASZLO is president of the Laszlo Institute of New Paradigm Research, founder and president of the Club of Budapest, fellow of the World Academy of Art and Science, member of the Hungarian Academy of Science and of the International Academy of Philosophy of Science, senator of the International Medici Academy, and editor of the periodical *World Futures: The Journal of New Paradigm Research.* He is the recipient of the Goi Peace Prize (2002), the International Mandir of Peace Prize (2005), the Conacreis Holistic Culture Prize (2009), the Ethics Prize of Milano (2014), and the Luxembourg World Peace Prize (2017). He was nominated for the Nobel Peace Prize in 2006 and 2007. He holds honorary Ph.D.'s from the United States, Canada, Finland, and Hungary.

Laszlo is the author of more than sixty books, translated into twenty-six languages, including *Reconnecting to the Source* (New York: St. Martin's Press, 2020).

About the Contributors

KINGSLEY L. DENNIS, **Ph.D.,** is a sociologist, researcher, and writer. He previously worked in the Sociology Department at Lancaster University, UK. Kingsley is the author of numerous articles on social futures; technology and new media communications; global affairs; and conscious evolution. He is also the author of over fifteen books, including *The Modern Seeker: A Perennial Psychology for Contemporary Times*; *Healing the Wounded Mind*; *Bardo Times*; *The Sacred Revival*; *The Phoenix Generation*; *New Consciousness for a New World*; *Struggle for Your Mind*; *After the Car*; and the celebrated *Dawn of the Akashic Age* (with Ervin Laszlo). Kingsley also runs his own publishing imprint, Beautiful Traitor Books (www.beautifultraitorbooks.com). For more information, visit his website at www.kingsleydennis.com.

MARIA SÁGI holds a Ph.D. in psychology at the Eotvos Lorand University of Budapest. Following seven years of intensive research in the psychology of music, she was awarded the C.Sc. degree of the Hungarian Academy of Sciences. Dr. Sági developed the information-medicine protocol pioneered by Austrian scientist Erich Koerbler into an encompassing method for diagnosing and treating human health problems. She is the author of twelve books and over one-hundred and fifty articles and research papers published in Hungarian, English, French, German, Italian, and Japanese, on topics as diverse as social and personality psychology, the psychology of music and art, in addition to information medicine, her principal field.

CHRISTOPHER M. BACHE, Ph.D., is professor emeritus in the department of Philosophy and Religious Studies at Youngstown State University where he taught for thirty-three years. He is also adjunct faculty at the California Institute of Integral Studies, Emeritus Fellow at the Institute of Noetic Sciences, and on the Advisory Council of Grof Legacy Training. Chris's passion has been the study of the philosophical implications of nonordinary states of consciousness, especially psychedelic states. An award-winning teacher and international speaker, Chris has written four books: *Lifecycles*—a study of reincarnation in light of contemporary consciousness research; *Dark Night, Early Dawn*—a pioneering work in psychedelic philosophy and collective consciousness; *The Living Classroom*, an exploration of collective fields of consciousness in teaching; and *LSD and the Mind of the Universe*, the story of his twenty-year journey with LSD.